Statistics and Pro

**Developed and Published
by
AIMS Education Foundation**

Research and Development

Sheldon Erickson
Betty Cordel
David Mitchell
Michelle Pauls
Dr. Richard Thiessen
Ann Wiebe
Dr. Arthur Wiebe
Jim Wilson

Illustrators

Brock Heasley
Renée Mason
David Schlotterback
Brenda Wood

Desktop Publisher

Tracey Lieder

Statistics and Probability

Developed and Published
by
AIMS Education Foundation

This book contains materials developed by the AIMS Education Foundation. **AIMS** (**A**ctivities **I**ntegrating **M**athematics and **S**cience) began in 1981 with a grant from the National Science Foundation. The non-profit AIMS Education Foundation publishes hands-on instructional materials that build conceptual understanding. The foundation also sponsors a national program of professional development through which educators may gain expertise in teaching math and science.

AIMS Education Foundation
1595 S. Chestnut Ave., Fresno, CA 93702 • 888.733.2467 • aimsedu.org

ISBN 978-1-60519-085-3

Printed in the United States of America

Statistics and Probability
Table of Contents

I Hear and I Forget,

I See and I Remember,

I Do and I Understand.

-Chinese Proverb

The Big Ideas of Statistics and Probability

This book addresses the big ideas of statistics and probability in an engaging, hands-on way that is appropriate for middle school students. In an age of information, students must become comfortable with dealing with data. They must understand how it is gathered, how to analyze the data to draw conclusions, and then how to use their generalizations to make predictions and choices. By being introduced to these ideas in engaging settings, students will be able to apply them in diverse situations.

Gathering Data—Statistics

To make generalizations about a group of things mathematically, one needs numbers that can be gathered by measuring or counting. This can be done by census or sample. A census counts or measures every member or item in a group. It is the most accurate way to make generalizations about a group since it contains all the data about the group. However, a census is often impractical because the group being surveyed is too large or too difficult to count. In this case, a sample of a group can be used to draw conclusions about the group.

For a sample to represent the group, it must be selected randomly. When conducting a survey, you cannot just ask your friends, for they are likely to share some of your preferences. Somehow, a representative slice of the group must be surveyed to gather accurate data. Random selection, or random sampling, provides this representation. In random sampling, every member of the group has an equal chance of being selected.

The size of a sample affects the accuracy of the generalizations made from the sample. Generally, the larger the sample, the more accurately it reflects the group; however, there are diminishing returns on accuracy of larger samples. A sample of 100 will not be twice as accurate as a sample of 50. In fact, doubling the sample size may only add a few percentage points of accuracy to the generalizations. It is for this reason that in a national survey of several thousand people, the pollsters are confident the generalizations are accurate within five percentage points. Surveying a million people would not change the accuracy enough to justify the magnitude of work and expense that would go into making such an exhaustive survey.

At an introductory level, students need to realize that a sample can provide a relatively accurate generalization of a much larger group if it is done randomly.

Measures of Central Tendency

In mathematics, *average* refers to the center or middle of a set of data. When someone refers to the "average American," no specific person is meant. Rather, the term indicates a composite person who represents the typical American with regards to habits, income, diet, family structure, health, etc. There are several ways to get at the average or center of the data.

Median

One way is to put the data in order and then find the value in the middle of the list. This is called the median. Like the median in a road, *median* means middle; it is a kind of average or center. The difficulty with this method when working with large amounts of data has traditionally been the tedious job of ordering the data. With the advent of computers, this job can be done almost instantaneously, greatly simplifying the process. The median is a good descriptor of average since it is not affected by unusually small or large extremes.

Mean

The more common measure of central tendency, what we often have in mind when we say "average," is called the *mean*. In the mean, the values are leveled by taking from the higher values and distributing to the lower values, which results in a leveled value of all the data. This is called the mean. The mean is calculated by finding the sum of all the values and dividing by the number of values. This method of averaging has been used historically because of the ease of the arithmetic. The mean is distorted by extremes and as a result does not describe the middle well when there are unusually high or low values.

Mode

A third measure of center is called the *mode*. This simply is the value that shows up the most often. The mode tells what is most common or typical in a set of data. Although it is often near the middle of the data, it maybe found much higher or lower than the center of data making it a less than reliable average.

When data are evenly distributed, the median, mean, and mode are nearly the same. As the distribution of data becomes uneven with unusually large or small data values, the averages may diverge.

Application

Students need to become familiar with the different types of averages for two reasons. First, they should be able to determine which average gives the best answer to the question or information about the situation at hand. If you were going to order skates to open a rink, you would be interested in the mode shoe size of the target population so you would have enough of the most common size on hand. If you were going to produce one-size-fits-all gloves, you would want to know the mean hand size of the target population so you would know where to center your dimensions. If you were an employer planning to create a new position at your business, you would want to know the median salary for similar positions at other companies in your area so that you could be competitive with your compensation package.

The second and probably more common need is to understand what the different averages really mean in a given situation. The same data can be used to support contradictory claims, and an understanding of the differences between mean and median sheds light on the apparent contradictions. For example, a given company might have a mean salary of $100,000 as a result of several $1,000,000 executive officers. That same company might have a median salary of $30,000 because of the large number of factory workers. In a dispute over wages, the lower-paid workers might site the mean salary as justification for a pay increase. The executives might argue that the median salary was reasonable when compared to industry standards.

Measures and Representations of Spread

Any form of an average gives a value that may or may not be very informative. The median will tell you the middle of the data. It will not tell you if the data are clumped very closely around this value or stretched out to great extremes. To be more informed, measures and displays of how the data are spread are needed.

Range

The simplest measure of spread is *range*. Range is the difference in value between the extremes of the data. The value of the range is used in formulas of advanced statistics, but is not very informative about the data itself. For students being introduced to statistics, the meaning of range in the context of its relationship to other data is more valuable than the calculated value of the range alone.

Consider the following scenario: A class of students has heights with a range of 30 cm. This value gives no information other than the fact that the tallest student is 30 cm taller than the shortest student. If the median height is given as 145 cm, the conclusion might be made that the heights in the class spread evenly and range from 130 cm to 160 cm. However, if the range is specified as being from 140 cm to 170 cm, a clearer picture emerges. The median's relationship to the extremes shows that the shorter half of the class varies in height by only 5 cm, while the taller half's heights range over 25 cm. The heights in the class are distributed much closer to 140 cm than 170 cm. The idea of range in relationship to measures of average informs the spread of the data.

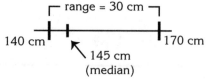

© 2012 AIMS Education Foundation

Quartiles and Extremes

The concept of median provides the foundation of another way to think about spread of the data. Once the data are sorted by value and the median is identified, the spread of the sorted data can be split into quarters. The value at each of these quarter splits is called a *quartile*. In a class of 32 students, height data could be divided into quarters with eight students falling in each quarter. The height of the shortest person would be the beginning and is called the lower extreme. The shortest quarter of people would consist of the eight shortest students. The first quarter division, called the lower quartile, would fall between the eighth and ninth students. Its value would be the mean of their heights. The lower quartile is also called the 25th percentile because one-fourth, or 25%, of the students are shorter than this height. The next quarter division is the median, and it falls between the 16th and 17th students. It can be referred to as the 50th percentile because half, or 50%, of the students are shorter than the median height and half are taller. The third quarter division, called the upper quartile, comes between the heights of the 24th and 25th students. This is the 75th percentile, with three-quarters, or 75%, of the students shorter than this height. The height of the tallest person is identified as the upper extreme.

Box-and-Whisker Plots

Distribution with extremes, quartiles, and a median is best represented with a box-and-whisker plot (sometimes called a box plot). A box-and-whisker plot is drawn in relationship to a scale that shows the range of the data. A box is drawn with the left side at the value of the lower quartile on the scale and the right side at the value of the upper quartile on the scale. A vertical line is drawn through the box at the median's value on the scale. This box represents the central half of the data. The lower and upper quarters are represented by horizontal lines, or whiskers, that connect the box to the extremes. Often, the central box spreads over less of the range's scale than either of the whiskers. This shows that the central half does not vary much from the median while the lower and upper quarters show a great variation.

Students being introduced to box-and-whisker plots often have difficulty interpreting them because they do not recognize that the size of the box and whiskers represents how the data are distributed along the range. They think each segment should be of equal length since they each represent one-fourth of the data values.

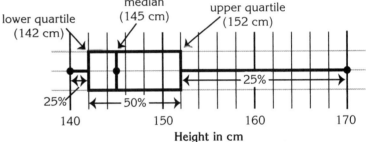

Correlation

Often it is necessary to determine if two data sets are related, as in the question: Is the number of hours spent watching television related to a student's grade point average? A graphic way to get an answer to this question is to gather both the numbers of hours watched and the GPAs of a group of students. Each student is represented by a point on graph with the number of hours of television viewing plotted along the x-axis and the GPA plotted along the y-axis. The pattern formed by the data points communicates the correlation of the data. Widely scattered points show no correlation. A tightly packed set of points clearly defining a line or curve shows a strong correlation. A more widely spread array of data points communicates a weaker correlation.

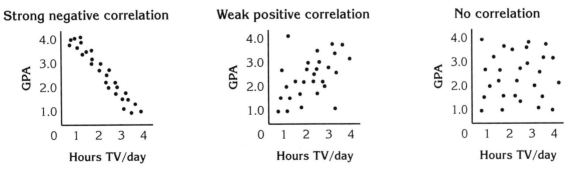

Studying correlation graphically provides an opportunity to see and make an application of algebra. Using a coordinate graph to see and interpret patterns is an emphasis of algebra. If the correlation forms a line, students have an opportunity to develop understanding of the concepts of linear equations.

Applying Data—Probability

While gathering and analyzing statistics at this introductory level tends to be descriptive in nature (defining a typical individual in the group and the spread of differences in the group), it also provides the opportunity to make predictions. If a large random survey of the heights of seventh graders has been made and a box-and-whisker plot has been generated, students can be expected to make predictions about the heights of other seventh graders. They should expect that out of four seventh graders, only one would be taller than the upper quartile value. Experiences like this allow for connections to the ways in which statistics are used to make predictions—from who is going to be the next president to whether it will rain tomorrow.

In probability, a prediction is always expressed as a number between zero and one. If an event has a probability of zero, it cannot happen. A probability of one means it must happen. In the previous example, one in four seventh graders were predicted to be taller than the upper quartile. Notice the ratio of one in four is equivalent to the fraction ¼ or to 25%. It is not likely that any given seventh grade student will be taller than the upper quartile. With probability expressed as a ratio, fraction, or percent, there is a wonderful opportunity to practice and apply proportional reasoning.

Simple games of chance provide situations that are easily studied and understood as well as being highly engaging. As students play the games, they gather data and develop a ratio of how many times they won compared to how many times they played. The fraction they get from experimenting expresses the probability they will win in the future. The more they play, the more accurately the fraction predicts the probability of winning in the future.

To keep from getting tricked into a game and not having much chance of winning, consider all the possible ways the game can turn out, and determine how many of those ways are winners. Considering the game theoretically before playing requires mathematical reasoning. To get at the probability, a ratio is needed. The number of ways to win compared to the total number of outcomes for the game must be compared. There is an obvious value in being able to deal with the probability of winning theoretically. Before playing, the chances of winning can be known or, in some cases, the game can be modified to increase the chances of winning.

As students become critical about how data is gathered, analyzed, and applied, they will more readily understand information they receive. By understanding the big ideas developed through the investigations in this book, they will be more critical of data and will be able to apply it more wisely in decisions they make.

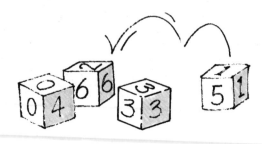

Topic
Data analysis: circle graph

Key Question
How can you use a circle graph to show the results of a survey?

Learning Goals
Students will:
- ask a question for use in a survey,
- conduct a survey and create a circle graph that displays the results, and
- determine the percentages for the responses.

Guiding Documents
Project 2061 Benchmarks
- *Organize information in simple tables and graphs and identify relationships they reveal.*
- *Find what percentage one number is of another and figure any percentage of any number.*
- *The graphic display of numbers may help to show patterns such as trends, varying rates of change, gaps, or clusters. Such patterns sometimes can be used to make predictions about the phenomena being graphed.*

Common Core State Standards for Mathematics*
- *Make sense of problems and persevere in solving them. (MP.1)*
- *Reason abstractly and quantitatively. (MP.2)*
- *Construct viable arguments and critique the reasoning of others. (MP.3)*
- *Attend to precision. (MP.6)*
- *Develop understanding of statistical variability. (6.SP.A)*

Math
Data analysis
   surveys
   data displays
     circle graph
Percentages

Integrated Processes
Observing
Communicating
Collecting and recording data
Interpreting data
Drawing conclusions

Materials
For each student:
   10 green 8-mm beads
   90 clear 8-mm beads
   1 m of crochet thread
   straight edge, such as a ruler
   2 *Percent Wheels* (see *Management 4*)

Background Information
Surveys and graphs are a part of students' everyday lives. The focus of this activity is for the students to gather and graph the results of a survey that they themselves conduct. They will construct a device, a percentage necklace, that will help them easily divide a circle graph into sectors that will show results of the survey they conducted. Students will also see how percents and circle graphs are related. They will be able to clearly see that 100 percent is the entire circle, the whole. The necklaces then allow them to determine the percentages for the various responses.

Management
1. The 8-mm beads can be purchased at craft stores or department stores. They are less expensive when purchased in large quantities.
2. The percentage necklace is constructed by stringing the 8-mm beads onto the crochet thread. The students will need to thread 9 clear beads for every 1 green bead. After all the beads are strung, the students will tie the ends of the string to form a necklace.

3. Direct the students to survey 10, 20, 25, or 50 people. These numbers will permit easy conversion when using the percentage necklace. The survey questions can connect to current events or things that are of high interest to the students. Encourage the students to select a survey topic that would give the opportunity for more than two responses. For example, finding out people's favorite sports team.

4. Copy the *Percent Wheels* on two different colors of card stock. Each student will need a wheel of each color.

Procedure

Part One
1. Ask the *Key Question* and state the first *Learning Goal*.
2. Tell the students that they will be conducting a survey. Challenge them to select a topic in which they are interested.
3. Distribute the first student page. Have them fill in the question that they will be asking. Direct the students to circle the number of people they will be surveying. Demonstrate, if necessary, how to keep ticks and tallies to show the responses.
4. Allow the students time to conduct their surveys.

Part Two
1. Ask the *Key Question* and state the second and third *Learning Goals*.
2. Distribute the beads and thread to each student and demonstrate how they are to construct the percentage necklace.
3. Ask the students to get their completed survey collection sheets. Check to make sure they have organized the data.
4. Distribute the blank circle graph page. Show the students how they can use the necklace to form a circle with the beads on the outside of the blank circle graph.
5. Direct the students to determine the value of each of the beads in the necklace as it relates to the circle and the number of people they surveyed. (If the students surveyed 10 people, it takes 10 beads (100/10) to represent each person. If they surveyed 50 people, it takes two beads (100/50) to represent each person.)
6. Point out the midpoint on the blank circle graph. Direct the students to draw one radius from the midpoint to the edge of the circle using a straight edge. Show them how to align one of the colored beads with this radius. This will be the starting point from which they will construct the circle graph. If students surveyed 10 people and they had three different responses, the graph would be divided into three different sectors. For example, the students could survey favorite types of ice cream. Say that three people surveyed responded vanilla, five people said chocolate, and two people said straw-berry. Three colored beads over from the starting radius a second radius would be drawn. A third radius would be drawn five additional

Favorite Ice Cream

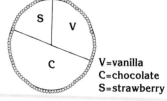

V=vanilla
C=chocolate
S=strawberry

colored beads over from the second radius. This divides the circle graph into three sections that show the results of this survey.
7. Allow time for students to construct their circle graphs for their surveys.

Part Three
1. Distribute the *Percent Wheels* to the students. Each student will need a wheel in each of the two colors. Invite them to cut out the circles and to cut along the dashed line to the midpoint of the circle.
2. Demonstrate how to insert one circle into the other.

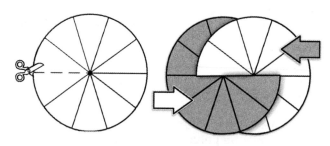

3. Ask students what they notice about the circles. [They are divided into 10 equal sectors. Each sector is divided into 10 smaller increments.]
4. Lead a discussion on how this circle is like the percent necklaces they made. [The small lines are the same as the clear beads of one color. The lines that go to the center are like the green beads.]
5. Have students turn the two circles so that 20 percent is showing in one color.
6. Invite students to call out percentages for the class to show on their wheels.
7. Have students position the *Percent Wheels* on the circle graphs to check the percentages of each sector.

Connecting Learning
1. What are the important parts of a circle graph?
2. How did the percentage necklace help with constructing the sectors of the circle graph?
3. How did you determine how many beads each response in the survey was worth?
4. How do circle graphs help us understand the results of a survey? [Circle graphs help us see part to whole relationships.]
5. Why is it important to know how many total people were surveyed when looking at the results of a circle graph? [A 30 percent sector that represents 300 people is not the same as a 30 percent sector that represents three people.]
6. How is the *Percent Wheel* like the percent necklace? How is it different?

7. If three-tenths of the class preferred chocolate ice cream and seven-tenths of the class preferred vanilla ice cream, what would your *Percent Wheel* look like to represent that data?

8. How are percents like fractions?

9. Did you have any problems with your question? Explain.

10. What other questions could you ask if you were doing another survey?

11. What kind of survey question would a scientist ask?

12. What are you wondering now?

Extension

Have students use an integrated graphing package or spreadsheet to collect, organize, and display their data.

Key Question

How can you use a circle graph to show the results of a survey?

Learning Goals

Students will:

- ask a question for use in a survey,
- create a circle graph that displays the results of the survey, and
- determine the percentages for the responses.

My question:

Number of people I will ask:

      10   20   25   50

                     (Circle one)

Collect and organize your data below.

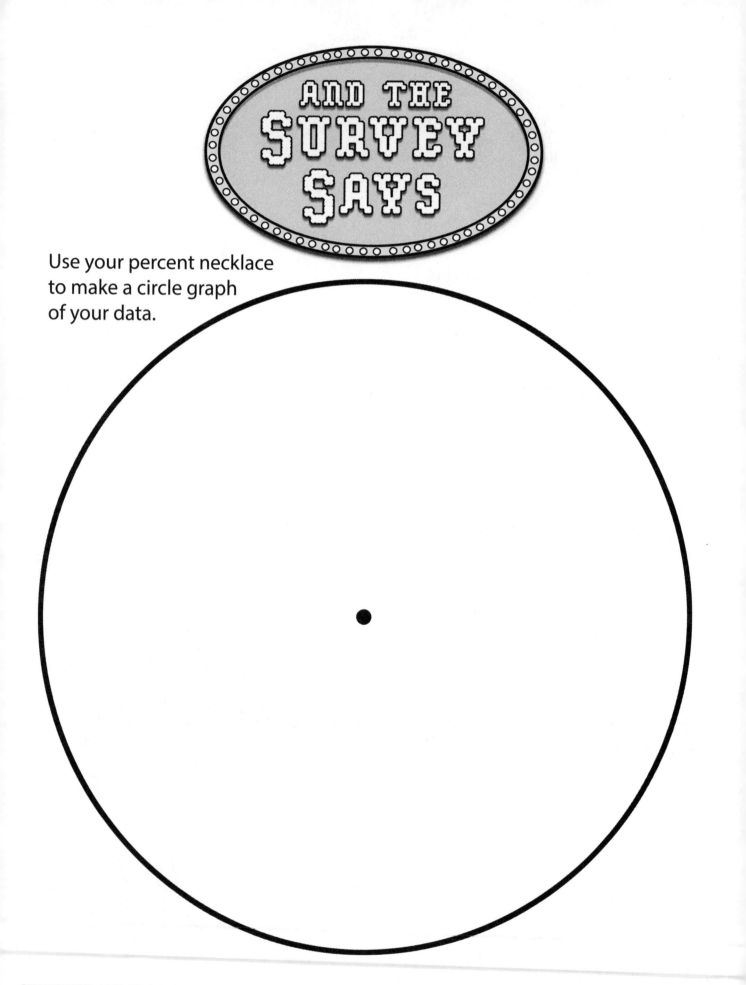

AND THE SURVEY SAYS

Use your percent necklace to make a circle graph of your data.

List four things your data tell you.

Write the percentages for your results.

Explain how the percentage necklace helped you solve these problems.

Percent Wheels

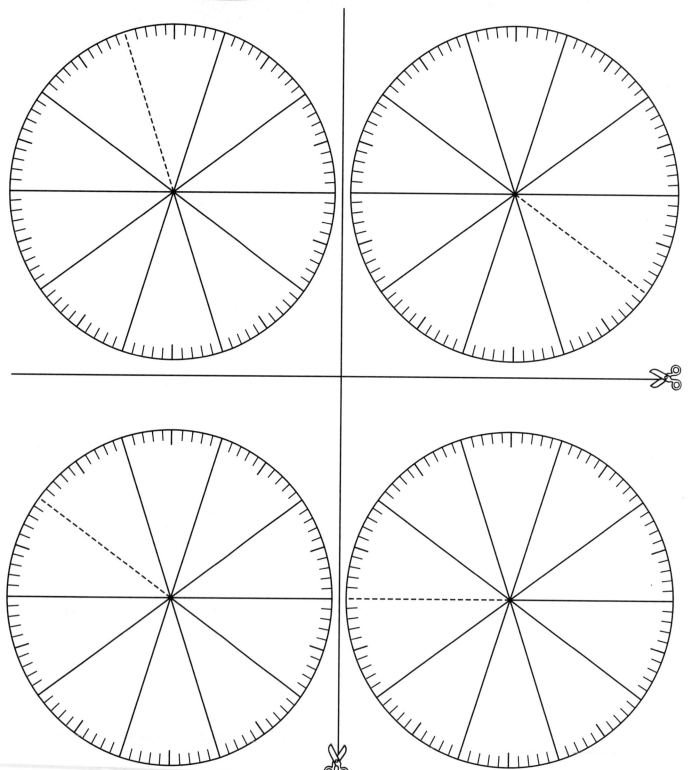

Connecting Learning

1. What question did you ask?

2. How did you gather your data? Did you have any problems?

3. Describe how you organized your data into a circle graph.

4. What conclusions did you draw from your survey?

5. How did the percentage necklace help with constructing the sectors of the circle graph?

6. How is the *Percent Wheel* like the percent necklace?

Connecting Learning

7. If three-tenths of the class preferred chocolate ice cream and seven-tenths of the class preferred vanilla ice cream, what would your *Percent Wheel* look like to represent that data?

8. How are percents like fractions?

9. Did you have any problems with your question? Explain.

10. What other questions could you ask if you were doing another survey?

11. What kind of survey question would a scientist ask?

12. What are you wondering now?

18

Getting to Know You

Topic
Organizing and displaying data

Key Question
How will you show your data?

Learning Goals
Students will:
- record data about themselves,
- find different people in the class who match each item, and
- graphically represent the data from a category.

Guiding Documents
Project 2061 Benchmark
- *Graphical display of numbers may make it possible to spot patterns that are not otherwise obvious, such as comparative size and trends.*

Common Core State Standards for Mathematics*
- *Make sense of problems and persevere in solving them. (MP.1)*
- *Reason abstractly and quantitatively. (MP.2)*
- *Construct viable arguments and critique the reasoning of others. (MP.3)*
- *Use appropriate tools strategically. (MP.5)*
- *Attend to precision. (MP.6)*
- *Develop understanding of statistical variability. (6.SP.A)*

Math
Data analysis
    data displays
        graphs

Integrated Processes
Observing
Collecting and recording data
Comparing and contrasting
Interpreting data

Materials
Scissors
Butcher paper
Marking pens
Crayons

Background Information
This activity is a way for you to assess the organizing and graphing skills of your students. This information will tell you whether you need to do further guided work with data organization or graphing before the class can comfortably move on to activities where graphing is expected.

Management
1. Since this activity lends itself to creative expression, you will not know what form the displays will take. Anticipate some possible material needs.
2. Consider doing this activity over two days. The first day students will write answers and play the matching game, after which the papers can be collected for you to peruse. The second day students can cut their categories apart, then illustrate the data for a category assigned to each group.
3. Do not tell students what they will be doing with the information they write. If they know they will be looking for matches in their class, they may purposely write answers that will match someone else's answers.
4. When looking for matches in which speed is an element, caution students about safety issues.
5. Labels naming the different categories can be placed around the room to designate collection areas.
6. A blank page is included for those who wish to devise their own categories for the survey.

Procedure
Day One
1. Tell students that they are going to do a survey that will help them learn about each other.
2. Distribute the survey and have each student write his or her answers in the second column. Purposely give no explanation about what they will do following this task as it may influence what they write.
3. Explain that students will be playing a matching game. They will have a limited time to find a different person to match each of their answers. In the third column, the matching person will sign or initial next to his or her answer, but can only do so one time on a given person's paper.

4. Give a starting signal. When most, but not necessarily all students, seem to be finished finding matches, give an ending signal.
5. Find who has the most matches and what some of those matches are. Collect their papers so you can read them and become better acquainted with the class.

Day Two
1. Decide which categories would be best (most informative, data without numerous groupings) for illustrating. The number of categories chosen should match the number of groups. List these so the class can see them.
2. Return the surveys to the class and have them cut across the horizontal lines so that the category and answer are in one piece. The matching initials should be cut away.

3. Designate specific areas in the room for each category and have students distribute their paper strips to the corresponding areas.

4. Explain that each group will receive one category of data and they are to create an appropriate graph or display (bar graph, circle graph, Venn diagram, dichotomous key, etc.) of that data. The data should be presented in a way that is easily understandable. Give groups the opportunity to pick the category with which they will work.
5. Distribute a large piece of paper to each group. Have each group discuss how they are going to make their display, then complete it.
6. Have the groups present their displays.

Connecting Learning
1. What was something new you learned about someone else?
2. What do you like about this data display?
3. In what other ways could this data have been displayed?
4. What do the data tell us about our class?
5. What categories do you think might show different results for people of different age groups? Why do you think this?
6. What questions would you like to use?
7. What are you wondering now?

Getting to Know You

Key Question

How will you show your data?

Learning Goals

Students will:

- record data about themselves,

- find different students in the class who match each item, and

- graphically represent the data from a category.

Getting to Know You

Favorite color		
Birthday month		
State or country of birth		
Eye color		
Favorite food		
Number of letters in first name		
Favorite sport		
Favorite pro-sports team		
Something you enjoy doing		
Kind of pet		
Favorite school subject		
Favorite TV show		
Number of years at this school		
Favorite holiday		

Getting to Know You

Getting to Know You

Connecting Learning

1. What was something new you learned about someone else?

2. What do you like about this data display?

3. In what other way could this data have been displayed?

4. What do the data tell us about our class?

5. What categories do you think might show different results for people of different age groups? Why do you think this?

6. What questions would you like to use?

7. What are you wondering now?

Counting Ch@r@cter$

VOLUME I, NUMBER III

Topic
Sampling

Key Question
How can you estimate how many characters of type (numbers, letters, symbols, punctuation) are on a page from the classified ads in a newspaper?

Learning Goals
Students will:
- use random sampling to estimate the number of characters on a page of classified ads, and
- use proportional reasoning to determine how many pages of classified ads it would take to see a million characters.

Guiding Documents
Project 2061 Benchmarks
- *Find the mean and median of a set of data.*
- *The mean, median, and mode tell different things about the middle of a data set.*
- *The larger a well-chosen sample is, the more accurately it is likely to represent the whole. but there are many ways of choosing a sample that can make it unrepresentative of the whole.*
- *Use calculators to compare amounts proportionally.*

Common Core State Standards for Mathematics*
- *Make sense of problems and persevere in solving them. (MP.1)*
- *Reason abstractly and quantitatively. (MP.2)*
- *Construct viable arguments and critique the reasoning of others. (MP.3)*
- *Use appropriate tools strategically. (MP.5)*
- *Attend to precision. (MP.6)*
- *Use random sampling to draw inferences about a population. (7.SP.A)*

Math
Estimation
Data collection
   sampling
Data analysis
   measures of central tendency
      mean, median, range
Proportional reasoning
   ratios
Problem solving

Integrated Processes
Observing
Predicting
Collecting and recording data
Interpreting data
Analyzing
Applying

Materials
For each group:
   one page of classified ads (see *Management 4*)
   yardstick or measuring tape
   sampling squares (see *Management 2*)

For each student:
   student pages
   calculator

Background Information
There are two principal ways of gathering quantitative data—by census or by sample. In a census, every organism, object, event, etc., is counted. Since it is usually impractical or impossible to count every element, the preferred technique is sampling. For example, rather than count all the grains of sand on the beach, you can count the number of grains in a small area. Once that value is known, you can multiply to estimate the number of grains of sand on the beach. One of the most frequently used methods of sampling is random sampling. In random sampling, each element has an equal chance of appearing in the sample.

Measures of central tendency, or averages, are used to clarify what is typical or normal in a set of data. The mean, median, and mode are all measures of average or typical. Although an average often refers to the arithmetic average or mean, when it is used in a report such as the "average American," it many refer to any of the measures of central tendency.

In normally distributed data, the mean, median, and mode will be the same. Statisticians generally regard the median as the most consistent measure. The mean is skewed by extremes and mode often depends on the precision of measuring. When deciding on a number to use as an "average," one must make a judgment call on which is most reliable in the situation.

Management

1. Students need to work in groups of two to six students.
2. Copy the page of sampling squares onto card stock. Each group will need six squares.
3. Six random samples is the minimum number required to get reliable data. If you choose to do more, each group will need more sampling squares.
4. Each group will need one page from the classified section of the newspaper. Try to select pages that are mostly text and don't have a lot of pictures or large ads.

Procedure

1. Ask the *Key Question* and state the *Learning Goals*.
2. Have students get into groups and distribute the classified pages. Ask students to estimate the number of characters on the page and to begin counting.
3. After a short period of time, students should realize that there must be an "easier way" to count the characters. Ask them to suggest some possible methods.
4. Distribute the student pages, yardsticks, and the sampling squares. Have the students record their individual estimates of the number of characters on their classified pages, then average their group's estimates.
5. Have students measure the length and width of the printed portion of the page and calculate its area.
6. Instruct groups to lay their newspapers flat on the floor. Tell them to select one person to stand about a foot from the edge of the paper and toss the sampling squares onto the page. If any squares miss the paper entirely, they should be dropped again until they are on the page.
7. Have them trace around the edge of each square with a pencil and remove the squares.
8. Tell students that they now need to count the number of characters in each square. Explain that everything counts (letters, symbols, punctuation marks, etc.). If half or more of a character falls within the square, it should be counted.
9. Have students complete the student pages, using their calculators as needed to do the calculations.

Connecting Learning

1. How did your group determine the typical number of characters in a square inch?
2. Did each group come up with the same number? Why or why not?
3. Why were you instructed to toss the sampling squares onto the newspaper?
4. How many characters did your group determine were on your page?

5. Did each group come up with the same number? Why or why not?
6. What are the mean, median, and range of characters on a page for the class?
7. How many characters would you say are typically on a page?
8. Do you think your group's estimation or the class's estimation is a better predictor of the number of characters on a typical page? Why?
9. How many pages do you need to look at to see a million characters?
10. Was using a random sample a good way to get an accurate estimate of what is on a page?
11. How else could you apply this random sampling method to count a large number of things?

Extensions

1. Have students sample newspaper issues and estimate the number of characters in a typical newspaper.
2. Have students develop a sampling method to determine the number of characters in a common book such as textbook or dictionary.

Counting Ch@r@cter$

VOLUME I, NUMBER III

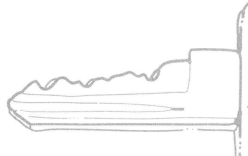

Key Question

How can you estimate how many characters of type (numbers, letters, symbols, punctuation) are on a page from the classified ads in a newspaper?

Learning Goals

Students will:

- use random sampling to estimate the number of characters on a page of classified ads, and

- use proportional reasoning to determine how many pages of classified ads it would take to see a million characters.

27

Counting Ch@r@cter$

VOLUME I, NUMBER III

Counting Ch@r@cter$
VOLUME I, NUMBER III

How many characters (numbers, letters, symbols, and punctuation) do you think are on this page of classified ads?

MY GUESS

DETERMINE PRINTED AREA

Measure and determine the area of the printed portion of your page.

_____in x _____in = _____in²
LENGTH   **WIDTH**   **AREA**

GATHER A RANDOM SAMPLE

1. Lay the classified ad page flat on the floor. While standing about 10 inches from the edge of the page, toss each of the six squares onto the printed portion of the page. All must land within the printed portion to be considered. If squares fall outside the printed portion, toss them again. Carefully trace an outline around each square.

2. Count the number of characters in each square. Where characters are split by the outline, they are counted only if half or more than half of the character lies within the square.

ANALYZE THE DATA

Record the number of characters per square and calculate the mean.

Square One	Square Two	Square Three	Square Four	Square Five	Square Six	Total	Mean

Sort the number of characters per square and determine the median and range.

Fewest Characters					Most Characters	Median	Range

Counting Ch@r@cter$

From the data, decide as a group how many characters you think are found in a typical square inch of print. Explain how you chose this number.

NUMBER OF CHARACTERS PER SQUARE INCH

USE DATA TO ESTIMATE

$$\underline{\hspace{4cm}} \times \underline{\hspace{4cm}} = \underline{\hspace{4cm}}$$

CHARACTER/in²　　　　　　in² /PAGES　　　　　　　CHARACTERS/PAGE

Share your data with the class to get a better estimation.

Group One	Group Two	Group Three	Group Four	Group Five	Group Six	Group Seven	Group Eight	Group Nine	Group Ten

Class Mean	
Class Median	
Class Range	

Using the data, what is a good estimation of the number of characters on a page of the classified ads?

APPLY THE DATA

How many pages of classified ads will it take to make a million characters?

Counting Ch@r@cter$

VOLUME I, NUMBER III

Connecting Learning

1. How did your group determine the typical number of characters in a square inch?

2. Did each group come up with the same number? Why or why not?

3. Why were you instructed to toss the sampling squares onto the newspaper?

4. How many characters did your group determine were on your page?

5. Did each group come up with the same number? Why or why not?

6. What are the mean, median, and range of characters on a page for the class?

Counting Ch@r@cter$

Connecting Learning

7. How many characters would you say are typically on a page?

8. Do you think your group's estimation or the class's estimation is a better predictor of the number of characters on a typical page? Why?

9. How many pages do you need to look at to see a million characters?

10. Was using a random sample a good way to get an accurate estimate of what is on a page?

11. How else could you apply this random sampling method to count a large number of things?

32

Topic
Concept of averages: median and mean

Key Question
What is the typical word length?

Learning Goals
Students will:
- make lists and line plots of the lengths of words to develop the concept of median, and
- cut and rearrange words to develop the concept of mean.

Guiding Documents
Project 2061 Benchmarks
- *Find the mean and median of a set of data.*
- *The mean, median, and mode tell different things about the middle of a data set.*
- *Spreading data out on a number line helps to see what the extremes are, where they pile up, and where the gaps are. A summary of data includes where the middle is and how much spread is around it.*

Common Core State Standards for Mathematics*
- *Make sense of problems and persevere in solving them. (MP.1)*
- *Reason abstractly and quantitatively. (MP.2)*
- *Construct viable arguments and critique the reasoning of others. (MP.3)*
- *Use appropriate tools strategically. (MP.5)*
- *Summarize and describe distributions. (6.SP.B)*

Math
Data analysis
  measures of central tendency
    median, mean
  data displays
    line plot
  spread
Ordering

Integrated Processes
Observing
Comparing and contrasting
Interpreting data

Materials
Dictionaries
1-cm graph paper or the grid provided
Scissors
Glue or tape
Lima beans or blocks

Background Information
Measures of Center

A key aspect of data is where it is *centered*. There are three measures of center (averages)—the mode, the median, and the mean. The *mode* is the value that occurs most frequently in a set of data. It can easily be skewed by data bunched on the low or high end.

The *median* is the middle value in an ordered set of data. If the set is odd, the median is the middle number. If the set is even, the median is the mean of the two middle numbers. Half of the numbers are at or above the median and half below.

The National Council of Teachers of Mathematics (NCTM) emphasizes the median for the middle elementary grades because it is easier to understand, easier to compute, and, in most cases, not affected by large or small values outside the bulk of the data.

The arithmetic *mean* is the average computed by adding all of the values (elements) in the set and dividing by the number of values in the set. In this activity, the concept of mean is introduced by "robbing the rich to give to the poor." Letters from longer words are given to shorter words until all the words in the set are as even as possible. It then progresses to the calculation.

Spread

Another key aspect of data is its *spread*. For small amounts of data, say less than 25, a line plot is a quick way to visually show the spread. A line plot, also called a frequency graph, is simply a number line with an *x* marked for each value in the data set. It reveals the largest and smallest values (range) and where there are clusters and gaps.

Example: 2, 3, 3, 7, 8, 12, 25

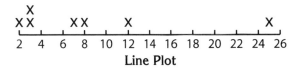
Line Plot

The mode is 3. The median is 7. The mean is 8.6.

This activity is more than an exercise in identifying the median (or mean). Students need to understand what the median tells them about the data and how the median relates to the spread. Because this understanding rarely comes from one experience and because NCTM stresses the importance of comparing related data sets, three sets of data are provided.

Management

1. This is a three-part investigation. In *Part One*, the blue, yellow, and red word webs are developed, forming the data sets used to explore line plots and the median. *Part Two* introduces a concrete way to determine the mean, while *Part Three* leads to calculating the mean.
2. Consider doing *Part One* the first day and *Parts Two* and *Three* the next day.
3. Organize students into pairs.

Procedure

Part One

1. Encourage student pairs to brainstorm words for blue. Conduct a sharing session, writing down their suggestions.
2. Explain that students will be working with synonyms for blue, yellow, and red. Some of their brainstormed words may be included, and new ones can be added to the class list.
3. Give student pairs the color web page and instruct them to follow the directions.

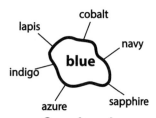

Sample web

4. After confirming the color groupings, tell students they will be using these word sets to learn more about interpreting data.
5. Distribute the median/line plot page.
6. Have students order each color word group and identify the median. For those who need a more concrete process, have them write the seven words on graph paper (one letter per box), cut out the word strips, order the strips, and record.
7. Model how to construct a line plot and direct students to complete the three line plots on the page.

blue
navy
azure
lapis (median-5)
cobalt
indigo
sapphire

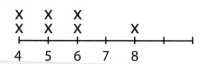

8. For each word group, ask students to describe how the word lengths are spread (including range) and where the median falls.

Part Two

1. Tell students they will now learn about another kind of center or average.
2. Give students the mean page and the graph paper.
3. Have students follow the directions on the page, focusing on one color word group at a time.

b	l	u	e	e
n	a	v	y	r
a	z	u	r	e
l	a	p	i	s
i	n	d	i	g	o
c	o	b	a	l	t
s	a	p	p	h	i	r	e

Starting edge

4. Ask students to describe the mean for each group. This is not a calculated mean, so students will talk in terms of whole numbers such as "It's almost halfway between five and six" or "It's a little more than six."
5. For each color word group, instruct students to compare the mean to the median and to the line plot in *Part One*.

Part Three

1. Introduce another medium for determining the mean by distributing lima beans or blocks to pairs of students.
2. Choose a group of words and have students count out one bean for each letter, pool the beans, and divide them into fair shares (seven groups for seven words).
3. Repeat, as needed, for other word groups.
4. Explain that, many times, the mean is somewhere between two whole numbers. Ask if there is another way to determine the mean more precisely. [Add the numbers and divide by the number of elements in the original set.]
5. Instruct students to calculate the mean for each group of words. Round to the nearest tenth.

Connecting Learning

Part One

1. Describe the spread for the blue words. [The range is 4 to 8. Most are 4s, 5s, and 6s; there is a gap before the 8.] Where does the median fall on the line plot? [The median is 5, on the lower end of the line plot.]
2. Describe the spread for the yellow words. [The range is 5 to 11. Most words are 5s and 6s; then there is a big gap and one 11.] Where does the median fall on the line plot? [The median is 6, in the area where most of the *x*s are clustered.]

3. Describe the spread for the red words. [The range is 3 to 8. The words are quite evenly spread.] Where does the median fall on the line plot? [The median, 6, really looks like it is in the middle.]
4. How does the spread on the three line plots compare? [They look different. The ranges vary. The yellow words are bunched up at 5 and 6 with a large gap while the red words are almost even all the way across. The blue words have a little bunching and a little gap.]
5. Make a line plot combining the words from all three groups. What is the range? [3 to 11] How can you find the median just by looking at the line plot? [For 21 pieces of data, the 11th one would be in the middle. Since the line plot is in order from lowest number to highest number, start counting at the low end until you reach the 11th x.] What is the median? [6, just barely]

Part Two
1. How would you describe the mean for the blue words? [about halfway between 5 and 6] Mark its approximate position on the line plot. How does it compare to the median? [The mean is higher than the median (5).]
2. Describe the mean for the yellow words. [more than 6] Mark its approximate position. How does it compare to the median? [The mean is slightly more than the median (6).]
3. Describe the mean for the red words. [about halfway between 5 and 6] Mark its approximate position. How does it compare to the median? [The mean is lower than the median (6).]

4. Is each mean represented by a value on the line plot? Explain. [No, each one happens to be between the whole numbers on the line.]
5. Can you think of another way to determine the mean length of these words? [Add the total number of letters in the set and divide by the number of words.]

Part Three
1. How can we determine the mean more precisely? [Add the number of letters and divide by the number of words in the set.]
2. Challenge: Calculate the mean of all 21 words (or a word set suggested in the *Extension*).

Extension/Assessment
To further develop or to assess data analysis skills, have students construct line plots and determine the median and mean for an odd number of:
- words for green or another color,
- names of car companies,
- first names of classmates, or
- NFL or NBA team names.

Centering on Color

Key Question

What is the typical word length?

Learning Goals

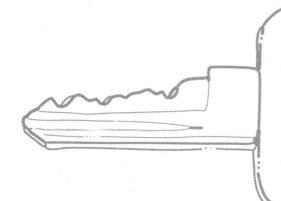

Students will:

- make lists and line graphs of the length of words to develop the concept of median, and
- cut and rearrange words to develop the concept of mean.

Build the blue, yellow, and red color webs using the synonyms below. Use what you know and a dictionary.

crimson	lemon	canary	ruby	cobalt	burgundy
navy	topaz	indigo	scarlet	buttercream	lapis
cherry	sapphire	brick	blond	azure	banana

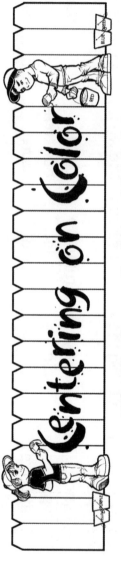

Centering on Color

What is the typical word length?

For each word group:
• Write the seven words in order from shortest to longest.
• Record the number of letters in the middle word. ⟶ This is the median, the center of an ordered list. It is one kind of average, one way of describing where the data are centered.
• Make a line plot of word lengths.

Blue

Yellow

Red

For each color group, describe how the word lengths are spread. Where is the median on each line plot?

Centering on Color

b	l	u	e											

Starting edge

y	e	l	l	o	w									

Starting edge

r	e	d												

Starting edge

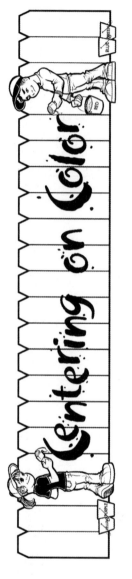

Centering on Color

What is the typical word length?

For each word group:

- Write the seven words in a grid, one letter per box.
- Cut around the outer edge of the whole group of words. Glue or tape down the starting edge only.
- Exchange just enough letters, by cutting, so that each word is the same length or only one letter different. Glue or tape.
- Count the number of letters each word has now. This is the mean, another kind of average or measure of center.

Blue

Yellow

Red

For each color group, describe the mean.

How does the mean compare to the median? How does the mean relate to the line plot?

Connecting Learning

Part One

1. Describe the spread for the blue words. Where does the median fall on the line plot?

2. Describe the spread for the yellow words. Where does the median fall on the line plot?

3. Describe the spread for the red words. Where does the median fall on the line plot?

4. How does the spread on the three line plots compare?

5. Make a line plot combining the words from all three groups. What is the range? How can you find the median just by looking at the line plot? What is the median?

Part Two

1. How would you describe the mean for the blue words? Mark its approximate position on the line plot. How does it compare to the median?

Connecting Learning

2. Describe the mean for the yellow words. Mark its approximate position. How does it compare to the median?

3. Describe the mean for the red words. Mark its approximate position. How does it compare to the median?

4. Is each mean represented by a value on the line plot? Explain.

5. Can you think of another way to determine the mean length of these words?

Part Three

1. How can we determine the mean more precisely?

2. Challenge: Calculate the mean of all 21 words.

The Marbelous Rolls

Topics
Data analysis: mean, median, and range
Scientific inquiry

Key Question
How does the distance a marble rolls down an inclined plane affect the distance it will roll on a carpet?

Learning Goals
Students will:
- identify the manipulated and responding variables in an experiment,
- form hypotheses, and
- carry out an experiment to test their hypotheses about the effect of the distance marbles roll down an inclined plane on the distance they roll on a carpet.

Guiding Documents
Project 2061 Benchmarks
- *Tables and graphs can show how values of one quantity are related to values of another.*
- *Use numerical data in describing and comparing objects and events.*
- *The mean, median, and mode tell different things about the middle of a data set.*
- *Find the mean and median of a set of data.*

NRC Standards
- *The motion of an object can be described by its position, direction of motion, and speed. That motion can be measured and represented on a graph.*
- *Identify questions that can be answered through scientific investigations.*
- *Design and conduct a scientific investigation.*
- *Use appropriate tools and techniques to gather, analyze, and interpret data.*
- *Use mathematics in all aspects of scientific inquiry.*

Common Core State Standards for Mathematics*
- *Make sense of problems and persevere in solving them. (MP.1)*
- *Reason abstractly and quantitatively. (MP.2)*
- *Construct viable arguments and critique the reasoning of others. (MP.3)*
- *Use appropriate tools strategically. (MP.5)*
- *Attend to precision. (MP.6)*
- *Summarize and describe distributions. (6.SP.B)*

Math
Measurement
    length
    angles
Data analysis
    measures of central tendency
        mean, median, range
    data displays
        line graph
Proportional reasoning
    ratios

Science
Physical science
Scientific inquiry
    variables
        manipulated, responding

Integrated Processes
Observing
Collecting and recording data
Comparing and contrasting
Interpreting data
Hypothesizing
Analyzing

Materials
For each group:
    2 meter sticks
    metric measuring tape
    transparent tape
    masking tape
    5 good quality marbles of uniform size and mass
    protractor
    colored pencils, optional

For each student:
    student pages

For the class:
    transparency of the graph page

Background Information
Gravitational potential energy is the energy an object has due to its position in a gravitational field. The marble sitting on the ramp has gravitational potential energy; it could move if it were not being held in place. The energy is not released until the marble is released. The higher the marble is positioned above the floor, the more potential energy it has.

As the marble rolls down the ramp, the potential energy is changed into kinetic energy. Kinetic energy is the energy an object has due to motion. The greater the marble's energy, the farther the marble will travel once it leaves the ramp. A change in height causes a change in the distance the marble will roll. By placing the marble higher off the floor, it therefore has more potential energy.

Generally, the steeper the slope, the farther the marble will roll. Eventually, the height of the slope will become so steep that the marble will not continue to roll as far as it did previously. This is due to the fact that the forward motion of the marble is absorbed into the floor.

Two statistical measures of central tendency are involved in this investigation, median and mean. Since five marbles are used in each trial, when the marbles come to rest, the marble in the middle will be the median distance. The average of the five distances will be the mean distance. The range is the distance between the centers of the marble rolling the farthest and the marble rolling the shortest distance.

When five marbles are rolled in a given test, it may be that one or two will strike other previously rolled marbles. While this may affect the range, it should have almost no effect on the mean or median. In studying the resolution of forces, we learn that the momentum lost by one marble is transferred to the marble it strikes and the combined effect is the same as if contact had not occurred.

Management
1. This activity works well with groups of four students, one to serve as recorder and coordinator; two as members of the measurement team, and the fourth as the marble launcher.
2. Construct the inclined plane by placing two meter sticks flat on the table next to each other with numbers matching. Cover the joint along its entire length with a single strip of masking tape. Next, create a v-shaped channel so the meter sticks form an angle just a little greater than 90 degrees. Use a 10-12 centimeter length of transparent tape across the 100 cm end to firmly hold the channel in position. Use additional strips as necessary to assure that the channel will withstand the necessary handling.

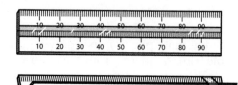

3. Set up the apparatus on a carpeted floor. If a carpeted floor is not available, indoor-outdoor carpeting is an economical alternative and can be purchased in most building supply stores. Eighteen inches of six-foot wide carpeting is sufficient for one setup. Use books or other objects to raise the meter stick channel to the height of about 15 centimeters at its upper (100 centimeter) end and place the lower (0 centimeter) end on the carpet.
4. All measurements should be made to the centers of the marbles. Also, the center of the marble is to be placed at the 15-centimeter mark on the inclined plane for the first test.

Procedure
1. Have students get into their groups and distribute the materials.
2. Guide groups through the ramp construction process as described in *Management*.
3. Explain that groups will be releasing five marbles from different heights and measuring how far they roll on the carpet. Describe the four roles (recorder, measurers, and marble launcher) and have students decide who will do each.
4. Have students identify the manipulated (independent) and responding (dependent) variables in the experiment. [The manipulated variable is the height from which the marble is released. The responding variable is the distance the marble rolls on the carpet.]
5. Instruct each group to form a hypothesis that includes these variables and to record it on the student page.
6. Have groups use their protractors to determine the angle at which the plane is inclined and record that on the student page.
7. Tell students that they will be rolling all five marbles before taking any measurements. Discuss the null effect on the median and mean in the event any marbles collide.
8. Have students practice putting the marble on the inclined plane and releasing it so that the releasing action does not introduce any distorting effect.
9. Once they have mastered this procedure, instruct them to roll their five marbles from a distance of 15 cm.
10. When all five marbles have been rolled, instruct them to measure and record the range and the distance each marble rolled from the base of the inclined plane.
11. Have students compute the ratio of the mean distance rolled over the distance rolled on the plane.
12. Instruct groups to repeat this process for each of the other distances (30 cm, 45 cm, 60 cm, and 75 cm).

13. Use a transparency of the graph page on the overhead projector to show students how you want them to record their data. Using one group's data as an example, mark the least and greatest rolls for each distance rolled on the inclined plane and connect these two points with a vertical line. This represents the range. You may want to use a different color for each distance rolled to help distinguish the data.

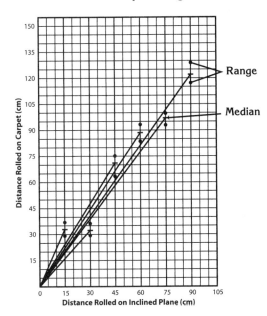

14. Mark the median roll for each distance traveled using a short horizontal line. Connect this median value to the origin (0, 0).

15. Give students time to graph the data from their groups. Have them study the relationship among the carpet/plane ratios as well as the ratios' relationships to the slope of the lines drawn on the graph. (If the ratios are in a relatively small range and the lines of the graph show similar slopes, there is an indication that the distance rolled on the carpet is directly proportional to the distance rolled on the inclined plane.)

16. Have students use all of this information to predict and record the mean and median distances the five marbles will roll when released from the 90 cm mark on the inclined plane.

17. Give time for groups to test their predictions and close with a final time of discussion and sharing.

Connecting Learning

1. What were the variables in this experiment? [the distance rolled on the inclined plane, the distance rolled on the carpet] Which was manipulated and which was responding? [The distance rolled on the inclined plane is the manipulated variable, and the distance rolled on the carpet is the responding variable.]

2. What was your group's hypothesis? Was your hypothesis correct? How do you know?

3. How do the ranges of the different distances compare? What might have caused any differences?

4. How do the medians and means compare? What does this tell you?

5. How does the carpet/plane ratio relate to the graph for that distance? [The ratio equals the slope (rise over run) of the graph for that distance.]

6. By looking at your graph, do you see any data that you think needs to be rechecked? Why do you think that?

7. Was your prediction of how far the marble would roll when released from 90 cm correct? Why or why not?

8. What are you wondering now?

The Marbelous Rolls

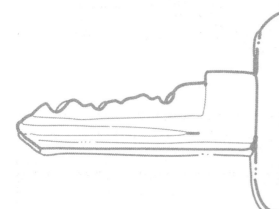

Key Question

How does the distance a marble rolls down an inclined plane affect the distance it will roll on a carpet?

Learning Goals

Students will:

- identify the manipulated and responding variables in an experiment,
- form hypotheses, and
- carry out an experiment to test their hypotheses about the effect of the distance marbles roll down an inclined plane on the distance they roll on a carpet.

The Marbelous Rolls

Our group's hypothesis is:

Angle of inclined plane: _____

Measure distances rolled on the carpet to the nearest 0.5 centimeter. Complete the table and compute the ratio: (mean distance rolled on the carpet)/(distance rolled on plane).

Distance rolled on an inclined plane	Distances rolled on carpet (cm) Marble					Range of distances (cm)	Median distance (cm)	Mean distance (cm)	Carpet Plane ratio
	A	B	C	D	E				
15 cm									
30 cm									
45 cm									
60 cm									
75 cm									
Prediction of the median and mean distances rolled on the carpet for 90 centimeters on the inclined plane.									
90 cm									

Our group observed:

The data lead us to accept/reject our hypothesis because:

The Marbelous Rolls

Graph your results. Plot the highest and lowest values for each distance rolled on the plane. This is the range. Mark the median distance traveled for each distance rolled on the plane with a horizontal line. Connect this line to the origin (0, 0).

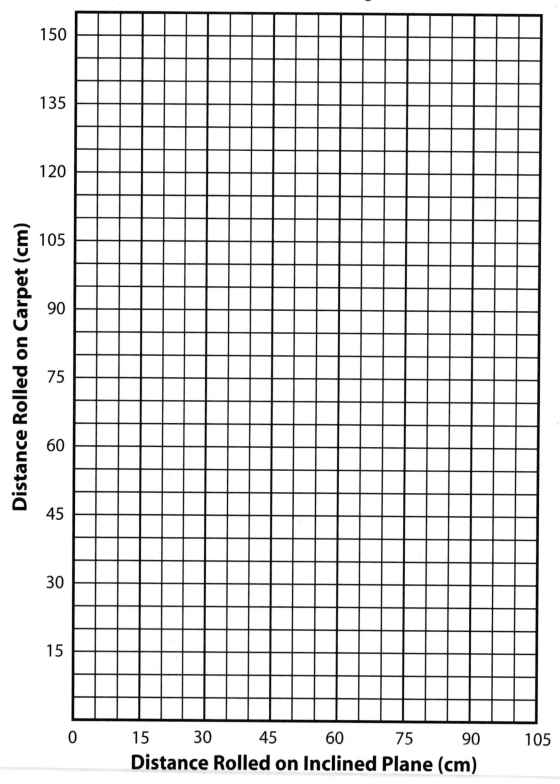

Distance Rolled on Carpet (cm) (y-axis): 15, 30, 45, 60, 75, 90, 105, 120, 135, 150

Distance Rolled on Inclined Plane (cm) (x-axis): 0, 15, 30, 45, 60, 75, 90, 105

The Marbelous Rolls

Connecting Learning

1. What were the variables in this experiment? Which was manipulated and which was responding?

2. What was your group's hypothesis? Was your hypothesis correct? How do you know?

3. How do the ranges of the different distances compare? What might have caused any differences?

4. How do the medians and means compare? What does this tell you?

The Marbelous Rolls

Connecting Learning

5. How does the carpet/plane ratio relate to the graph for that distance?

6. By looking at your graph, do you see any data that you think needs to be rechecked? Why do you think that?

7. Was your prediction of how far the marble would roll when released from 90 cm correct? Why or why not?

8. What are you wondering now?

Topic
Stem-and-leaf plots

Key Question
What will a stem-and-leaf plot tell us about the temperatures of cities in the Northern Hemisphere and cities in the Southern Hemisphere?

Learning Goals
Students will:
- plot temperature data into stem-and-leaf plots, and
- make back-to-back plots to compare ranges of temperatures.

Guiding Documents
Project 2061 Benchmarks
- *The meaning of numerals in many-digit numbers depends on their positions.*
- *Use numerical data in describing and comparing objects and events.*
- *Mathematics is the study of many kinds of patterns, including numbers and shapes and operations on them. Sometimes patterns are studied because they help to explain how the world works or how to solve practical problems, sometimes because they are interesting in themselves.*
- *Mathematical ideas can be represented concretely, graphically, and symbolically.*
- *Because the earth turns daily on an axis that is tilted relative to the plane of the earth's yearly orbit around the sun, sunlight falls more intensely on different parts of the earth during the year. The difference in heating of the earth's surface produces the planet's seasons and weather patterns.*

NRC Standards
- *The sun is the major source of energy for phenomena on the earth's surface, such as growth of plants, winds, ocean currents, and the water cycle. Seasons result from variations in the amount of the sun's energy hitting the surface, due to the tilt of the earth's rotation on its axis and the length of the day.*
- *Weather changes from day to day and over the seasons. Weather can be described by measurable quantities, such as temperature, wind direction and speed, and precipitation.*

Common Core State Standards for Mathematics*
- *Make sense of problems and persevere in solving them. (MP.1)*
- *Reason abstractly and quantitatively. (MP.2)*
- *Construct viable arguments and critique the reasoning of others. (MP.3)*
- *Model with mathematics. (MP.4)*
- *Use appropriate tools strategically. (MP.5)*
- *Summarize and describe distributions. (6.SP.B)*

Math
Data analysis
    measures of central tendency
        mean, range
    data displays
        stem-and-leaf plot

Science
Earth science
    temperatures

Integrated Processes
Observing
Organizing data
Interpreting data
Comparing and contrasting
Generalizing

Materials
Sticky notes, 3" x 3", three colors
World map
Colored pencils, two colors
Data table (see *Management 1*)
Student pages

Background Information
A stem-and-leaf plot uses numerical data in a display similar to a bar graph. In an item of numerical data, the more significant units are referred to as the *stems;* the less significant units are referred to as the *leaves.* Consider the following example of data that clusters in the one hundreds: In the number 125, the 12 would be the stem and the 5 would be the leaf. The range of stems found in the data are listed in a column. Then each leaf of the data is placed alongside the appropriate stem. Each leaf occupies the same amount of space so that frequency comparisons can be made of the lengths of the rows of leaves—just like a bar graph. Stems with more leaves in their row have more data occurring in that range. The differing lengths of

rows of leaves make it very clear how the data are distributed and where most of the data are centered.

Teaching students the organizational skills used in making stem-and-leaf plots involves the discussion of place value and ranges. Several situations should be modeled in which data are collected and arranged in stem-and-leaf display before beginning this activity. If students do not understand the range of numbers found in each stem, the numbers can be ordered on index cards. The cards can be taped together and then accordion-folded. Label the top card with the stem, then unfold the cards to illustrated all the numbers found within that stem's range. For example: Four as a stem representing all the forties would include 40, 41, 42, 43, 44, 45, 46, 47, 48, and 49, which are written on the index cards. In the folded position, only the 4, the stem, would be exposed. Cross out all the fours in the ten's position to indicate that they are not written on the leaves.

Ask the students why they think the entire number is not written in the leaf. [It would be redundant, inefficient, to keep repeating the 4 because it is common to all the leaves.]

Data **can** be arranged in an *ordered* stem-and-leaf plot. The ordering of the leaves occurs after all the data have been recorded. At this time the leaves in each stem are ordered from least to greatest. Using 4 as a stem and leaves of 5, 0, 4, 0, 5, 9, 9, 7, 9, 1, 2, 2, 4, the data would be ordered as 0, 0, 1, 2, 2, 4, 4, 5, 5, 7, 9, 9, 9. Notice that all leaves are recorded. The ordered display helps students to easily determine which leaves occur most (or least) frequently.

Two stem-and-leaf plots can be arranged back to back. There are a couple of options for doing this in *Lots of Temperature Plots*. One is to compare the temperatures for a given month in the Northern Hemisphere with the same month's temperatures in the Southern Hemisphere. Another way would be to compare the two different month's temperatures within the same hemisphere. These options allow students to easily compare the ranges of temperatures for the data displayed.

This activity utilizes real-world temperature data for cities in both the Northern Hemisphere and the Southern Hemisphere. It should help students to see the effects of the Earth's tilted axis and the resulting seasonal differences. When the Earth is positioned so that the sun's rays shine more directly on the Northern Hemisphere (summer), the Southern Hemisphere is receiving less direct rays creating more winter-like conditions.

Management

1. The data table that is provided is optional. You can have students collect data from the Internet. However, it will be necessary to have data from both hemispheres and data from January, April, July, and October.

2. Temperatures for seven days per month have been given in the data table. The temperatures for those seven days should be averaged and used in the stem-and-leaf plots. This table will most likely appear overwhelming to the students. Once they start analyzing the information, however, they will find that the data make sense and are very workable.

3. This activity will take several periods to complete, especially if data are plotted for all four months and for both hemispheres. It is essential that data be gathered for two different months in the same hemisphere or for the same month in both hemispheres.

4. Students should work in small groups when constructing the single stem-and-leaf plots. They can work individually when constructing the back to back plots.

5. Make two transparencies of the class data sheet. One will be used to record the average monthly temperatures of cities in the Northern Hemisphere and the other for Southern Hemisphere.

6. One strategy for assigning cities to students would be to write numbers on slips of paper to correspond to the numbers on the data sheet and have each student draw a number. The cities need to be divided into two groups as to whether they are in the Northern or Southern Hemisphere.

7. If you do not want your students to have to deal with negative numbers, eliminate Fairbanks from the cities that are to be selected.

Procedure

1. Ask students what is meant by Northern Hemisphere (Southern Hemisphere). [That part of the Earth that is found north (south) of the Equator. Hemi = half.]

2. Inquire as to what they know about the temperatures in the Northern Hemisphere (Southern Hemisphere) during the month of _____ (January, April, July, or October). Be sure to let students discuss the variations of temperatures due to the differences in latitude.

3. Distribute the data sheet. Ask students to determine the cities that are in the Northern Hemisphere. If necessary, use a world map to find the cities' locations above or below the Equator. Be certain to check the answers in class so all students will be working with the same set of data.

4. Have the students use one colored pencil to lightly shade the cities and temperatures in the Northern Hemisphere and the other colored pencil to do the same to the cities and temperatures in the Southern Hemisphere.

5. Group the students (see *Management 4*) and have each group draw a number to indicate the city in the Northern Hemisphere (Southern Hemisphere) for which they are responsible.

6. Direct students to look at the dates on the data table. Ask what they notice about the number of dates. [There is a week's worth of data (seven entries) per month for each city.] Notify the students that they will need to find the average temperature for each week in each of the four months for their given cities.

7. Have the class choose a month to use to make the first stem-and-leaf plot. Invite each group of students to share the average temperature for their city during that month while another student writes them on the overhead transparency.

8. Once all the data are collected, have the students determine the range of temperatures in order to establish the stems for the stem-and-leaf plot.

9. Use one color of the sticky notes to record the stems. Stick the stems to the board or to a bulletin board. Make a label that says *Northern Hemisphere (or Southern Hemisphere), _____* (month).

10. Distribute a second color of the sticky notes for groups to record the leaves. Invite the students to add the leaves to the stems.

11. Take time to analyze the data: Which stem has the greatest number of leaves? What does this mean? What is the lowest temperature? What is the highest temperature?

12. Optional: Take the time to order the stem-and-leaf plot. Then, ask the students what things are easier to determine now that the data are ordered. [One thing that may be more evident is the mode temperature.]

13. Have students transfer the data to the activity page.

14. Continue the investigation by selecting another month for the same hemisphere or the other hemisphere and the same month as before. Use the third color of sticky notes for the leaves on this plot. Again, have students transfer the data to the activity page.

15. This same procedure can be continued for as long as desired through all months and both hemispheres. As a conclusion, have students construct and record a back-to-back, ordered stem-and-leaf plot from the data already collected in the single stem and leaf plots.

Connecting Learning

1. Of what other type of graph does a stem-and-leaf plot remind you? [a bar graph]

2. What information does a stem-and-leaf plot provide? [It shows frequencies or counts.]

3. What makes the stem-and-leaf plot efficient? [The leaves don't repeat the digit(s) of the stems.]

4. What were the temperature ranges on each of the stem-and-leaf plots? When making the back to back plot, did the stem have to be altered? Explain. (perhaps, especially if plotting January and July temperatures or if Northern Hemisphere temperatures were plotted with Southern Hemisphere temperatures)

5. What, if any, were the patterns in the data that were displayed on the stem-and-leaf plots?

6. From the data shown on the stem-and-leaf plots, what connections can you make to seasonal changes?

7. What interesting things did you find in the data table? [Some cities, such as, Nairobi, Kenya; Miami, FL; Mexico City, Mexico; Jakarta, Indonesia; and Honolulu, HI have temperatures that fluctuate very little throughout the year. Fairbanks, AK had temperatures ranging from 80 degrees to -9 degrees. Winnipeg, Canada's temperatures ranged from 82 degrees to -13 degrees.]

8. Do you think finding an average was a good way to report the temperatures? Explain. [Yes, averaging evens out the highs and the lows. If a daily temperature were used instead, that temperature might not be "normal" for that time of year; it might be the result of an unusual warm (or cold) front.] What would be even better than averaging seven days worth of temperatures? [Averaging the temperatures of all the days in a month, or better yet, all the days within a season.]

9. What other things would you like to investigate dealing with temperatures?

10. In what other situations would a stem-and-leaf plot be useful for observing data? [Example: looking at boys' heights as compared to girls' heights]

Extensions

1. Maintain a stem-and-leaf plot to record daily temperatures for cities in the United States.

2. When percentage grades are given, have students keep track of their percentages using a stem-and-leaf plot.

Key Question

What will a stem-and-leaf plot tell us about the temperatures of cities in the Northern Hemisphere and cities in the Southern Hemisphere?

Learning Goals

Students will:

- plot temperature data into stem-and-leaf plots, and
- make back-to-back plots to compare ranges of temperatures.

Daily High Temperatures—Degrees Fahrenheit

City	Jan 16	17	18	19	20	21	22	Apr 17	18	19	20	21	22	23	Jul 17	18	19	20	21	22	23	Oct 16	17	18	19	20	21	22
1. Albuquerque, NM	54	63	58	59	48	51	53	82	83	78	79	80	78	70	95	95	97	97	96	94	97	83	76	68	68	77	76	76
2. Atlanta, GA	51	46	44	46	45	39	41	73	78	84	77	81	68	77	86	91	95	95	93	95	93	82	84	81	66	55	62	66
3. Auckland, New Zealand	77	78	75	75	73	78	69	69	60	62	66	68	66	69	59	59	57	57	60	60	60	62	64	60	57	60	66	66
4. Beijing, China	28	28	30	32	35	33	41	66	69	73	71	62	68	68	82	84	78	84	84	84	91	73	66	68	64	62	68	64
5. Billings, MT	50	49	44	30	38	43	41	54	39	38	52	60	36	54	91	97	98	82	86	83	82	52	57	58	61	68	60	66
6. Boise, ID	47	55	45	40	39	41	43	61	55	52	54	50	52	59	100	97	85	82	88	82	92	59	61	67	63	67	67	68
7. Boston, MA	35	21	40	39	32	30	26	63	61	52	46	56	50	55	93	86	89	91	97	103	92	66	66	68	59	69	64	63
8. Buenos Aires, Argentina	87	78	82	78	77	80	84	68	71	69	68	62	71	66	57	53	53	50	55	60	59	73	62	66	68	75	77	69
9. Cairo, Egypt	66	60	64	66	68	69	68	100	98	78	75	69	75	77	95	96	100	96	100	93	93	82	84	91	78	77	80	78
10. Caracas, Venezuela	86	87	91	91	8	86	91	84	84	86	86	80	71	86	87	89	89	84	89	88	84	87	84	86	84	84	82	82
11. Chicago, IL	23	35	34	24	22	10	19	52	43	41	44	48	57	64	95	93	91	100	100	100	87	60	57	54	48	46	54	66
12. Cincinnati, OH	32	39	44	36	32	21	25	68	73	78	77	57	63	69	91	93	96	98	99	97	93	81	66	63	52	46	60	62
13. Dakar, Senegal	78	87	78	77	78	78	78	77	77	77	75	75	75	75	86	87	89	86	84	84	84	93	91	95	98	91	93	89
14. Dallas-Ft. Worth, TX	46	57	55	54	52	43	59	82	91	91	72	86	91	89	100	102	100	100	102	100	100	87	89	68	69	72	82	84
15. Detroit, MI	19	32	37	25	23	21	16	46	37	39	45	51	46	73	93	96	90	96	100	95	91	59	61	55	51	52	48	57
16. Fairbanks, AK	15	1	-20	-25	-31	-13	0	0	41	42	37	39	46	47	47	60	62	75	77	80	67	27	32	30	26	28	28	23
17. Honolulu, HI	79	79	75	90	84	83	82	82	82	84	83	85	87	86	87	87	87	87	87	87	87	87	86	86	88	86	86	89
18. Jakarta, Indonesia	80	80	82	86	86	82	84	87	91	89	89	86	91	93	89	93	87	91	87	86	86	95	91	82	95	89	93	89
19. Johannesburg, South Africa	71	75	84	78	78	75	77	66	66	66	64	69	64	62	60	62	64	66	68	68	66	78	75	73	75	68	71	78
20. Kansas City, KS	25	41	36	21	25	28	3	72	68	51	57	61	64	61	97	98	99	100	100	102	102	72	55	55	51	55	71	73
21. Lima, Peru	77	75	80	78	77	77	77	75	73	73	75	75	75	75	66	64	64	64	66	64	64	68	68	66	64	68	71	71
22. Madrid, Spain	57	42	53	51	53	44	44	71	73	69	68	68	57	62	87	87	82	89	95	89	89	80	73	75	77	66	69	73
23. Mexico City, Mexico	68	73	75	76	74	54		57	76	74	63	63	65		58	76	79	80	79	77	70	53	76	73	72	77	75	73
24. Miami, FL	76	73	83	85	82	83	72	89	87	86	86	85	85	85	97	93	90	95	93	93	93	82	79	82	82	78	77	81
25. Minneapolis-St. Paul, MN	15	25	19	13	9	3	6	49	50	45	46	50	46	48	93	98	97	96	86	89	85	59	58	47	49	55	61	61
26. Moscow, Russia	19	10	12	14	14	12	17	48	53	44	42	51	55	60	77	82	84	86	86	84	89	41	42	48	41	53	48	44
27. Nairobi, Kenya	82	83	67	81	63	81	77	72	72	72	72	73	73	73	77	78	73	77	71	75	78	68	75	75	71	68	78	78
28. New Delhi, India	62	66	69	66	68	71	71	95	91	93	93	95	95	96	91	89	95	95	93	89	89	93	91	91	89	91	87	89
29. New York City, NY	37	27	41	41	35	32	24	61	63	53	65	63	51	61	91	95	94	89	97	104	100	66	65	68	62	67	56	60
30. Paris, France	55	50	48	39	41	33	39	64	71	75	75	77	75	78	66	66	62	64	68	66	64	64	62	59	55	51	53	51
31. Rio de Janeiro, Brazil	84	91	89	89	91	87	86	84	84	84	86	86	87	89	80	75	73	78	82	77	73	71	71	75	69	75	75	71
32. Rome, Italy	60	60	57	51	51	48	44	68	69	69	71	69	69	68	87	86	82	78	80	80	78	73	68	71	66	68	64	64
33. Sacramento, CA	53	51	60	63	60	63	71	76	67	73	73	73	69	67	80	84	88	96	97	93	89	77	87	86	76	81	82	83
34. Stockholm, Sweden	41	41	37	33	33	30	28	60	60	55	60	68	60	62	73	71	75	75	64	82	73	51	44	50	48	46	46	55
35. Sydney, Australia	82	93	80	80	82	87	74	69	73	77	68	78	77	64	59	66	59	62	55	57	59	80	68	71	77	82	86	84
36. Tokyo, Japan	44	51	50	48	46	46	50	60	62	66	66	59	64	66	93	91	84	87	73	75	78	86	75	69	62	68	68	73
37. Winnipeg, Canada	6	6	-9	3	-2	-7	-9	35	39	42	48	51	46	55	89	91	93	87	84	68	68	50	44	42	41	48	53	57

City _____   Hemisphere _____

_____   _____   _____   _____
Month   Month   Month   Month

Temperatures

_____
_____
_____
_____
_____
_____
_____

Average [ ]   Average [ ]   Average [ ]   Average [ ]

✁

Lots of Temperature Plots

City _____   Hemisphere _____

_____   _____   _____   _____
Month   Month   Month   Month

Temperatures

_____
_____
_____
_____
_____
_____
_____

Average [ ]   Average [ ]   Average [ ]   Average [ ]

Class Data Sheet

City	Average Temperature (°F)			
	January	April	July	October

Stem-and-Leaf Plots

Fill in the stems. Record the data from the class. What do the data tell you?

Hemisphere: _____

Month: _____

Hemisphere: _____

Month: _____

Lots of Temperature Plots

Fill in the stems. Record the data from the class.

Stem-and-Leaf Plot

Hemisphere: _____   Hemisphere: _____

Month: _____   Month: _____

Compare both sides of the stem-and-leaf plot.
List three things the data tell you.

Connecting Learning

1. Of what other type of graph does a stem-and-leaf plot remind you?

2. What information does a stem-and-leaf plot provide?

3. What makes the stem-and-leaf plot efficient?

4. What were the temperature ranges on each of the stem-and-leaf plots? When making the back to back plot, did the stem have to be altered? Explain.

5. What, if any, were the patterns in the data that were displayed on the stem-and-leaf plots?

6. From the data shown on the stem-and-leaf plots, what connections can you make to seasonal changes?

Connecting Learning

7. What interesting things did you find in the data table?

8. Do you think finding an average was a good way to report the seasonal temperatures? Explain. What would be even better than averaging seven days worth of temperatures?

9. What other things would you like to investigate dealing with temperatures and seasons?

10. In what other situations would a stem-and-leaf plot be useful for observing data?

SORTS OF PENNIES

Topic
Central tendency and spread

Key Question
In what year was a typical penny currently in circulation minted, and how could you determine that year?

Learning Goals
Students will:
- organize a sample of pennies by minting date;
- determine measures of central tendency: mean, median, mode, and range; and
- construct a line plot and a box-and-whisker plot from the data to study the spread of data.

Guiding Documents
Project 2061 Benchmarks
- *Find the mean and median of a set of data.*
- *The mean, median, and mode tell different things about the middle of a data set.*
- *Comparison of data from two groups should involve comparing both their middles and the spreads around them.*
- *The larger a well-chosen sample is, the more accurately it is likely to represent the whole. but there are many ways of choosing a sample that can make it unrepresentative of the whole.*

Common Core State Standards for Mathematics*
- *Make sense of problems and persevere in solving them. (MP.1)*
- *Reason abstractly and quantitatively. (MP.2)*
- *Construct viable arguments and critique the reasoning of others. (MP.3)*
- *Model with mathematics. (MP.4)*
- *Use appropriate tools strategically. (MP.5)*
- *Attend to precision. (MP.6)*
- *Summarize and describe distributions. (6.SP.B)*

Math
Data analysis
   measures of central tendency
      mean, median, mode, range
   data displays
      line plot, box-and-whisker plot
   spread

Integrated Processes
Observing
Predicting
Collecting and recording data
Interpreting data
Analyzing
Applying

Materials
For each group:
   1 roll of pennies, current circulation

For each student:
   student pages

Background Information
Typically, pennies remain in circulation for about 30 years. The bulk of the pennies in circulation are no more than 10 years old. The typical age of a penny changes due to the number of pennies minted in a year but generally is between five and 10 years.

There are two principal ways of gathering quantitative data—by census or by sample. In a census, every organism, object, event, etc., is counted. Since it is usually impractical or impossible to count every element, the preferred technique is sampling. To study pennies, a random sample must be obtained. A roll of pennies from current circulation provides such a sample. If the pennies are taken from a collection that has been gathering in jar, they will have aged and will not provide current information.

A line plot provides a way to organize data to study central tendency and spread. Line plots are usually used when there is one group of data and fewer than 50 values. A line plot consists of a horizontal number line on which each value of a set is denoted by an *x* over the corresponding value on the number line. The number of *x*s above each score indicates how many times each score occurred.

A real line plot can be made with pennies by sequencing them by years and stacking those of the same year in a pile. The number pennies in each pile can be transferred to the plot as *x*s. By looking at the plot, one can easily identify the mode as the tallest stack. The median is found by determining the stack in which the middle coin is found. With 50 pennies, the middle is between the 25th and 26th coins. The general bulk of coins are found in the most recent years, with the coins becoming more scarce and spread out as the pennies age.

To provide a better description of the spread of the data, a listing of the minting date of each penny can be made. The high and low (extremes), middle (median), and quarter divisions (quartiles) can be easily identified. Making a box-and-whisker plot provides a visual display of the data from which applications can be made.

Management
1. Currently circulating coins are available in rolls of 50 pennies at banks.
2. The investigation is most effective if each group of four to six students has a roll of coins. It can be done as a class if small samples of coins are given to each group and then, as dates are identified, the pennies can be put into one real line plot in front of the class.

Procedure
1. Ask the *Key Question* and have students share their predictions and suggest ways to determine the typical minting year of a penny currently in circulation.
2. Distribute the pennies and instruct the students to determine the minting date, sequence them by years and stack those of the same year in a pile.
3. Have the students transfer the data from the real graph to the line plot on their record sheets.
4. Referring to the line plot, have students identify the mode, median, range, and generalizations they can make about pennies.
5. Using the line plot, have students record the minting date of each penny in sequence.
6. Instruct students to identify the median, extremes, and quartiles and use the measures to construct a box-and-whisker plot.
7. Have the groups compare their plots and discuss reasons for similarities and differences.
8. Referring to their box-and-whisker plots, have students predict the probability of finding a coin dated to different time periods.

Connecting Learning
Measures of Central Tendency
1. What is the mode of the minting dates of your pennies? ...median? ...range?
2. How do you see it on the line plot?
3. What does it tell you about pennies currently in circulation?

Line Plots
1. What patterns do you see in the heights on the line plot? [highest in recent years]
2. What does that tell you about these pennies? [Most are new.]
3. How does you data compare to the data of other groups? (General trends should be the same. Details may vary.)

Box-and-Whisker Plot
1. How much of the box-and-whisker plot is made of just the box? [just a small fraction of the length]
2. What fraction of the pennies does the box represent? [½ or 50%]
3. Why is half the data filling much less than half the plot? [Half of the pennies were minted in very few years.]
4. How do the two whiskers of the plot compare in length? (The left whisker should be significantly longer than the right whisker.)
5. What fraction of the pennies does each whisker represent? [Each whisker represents ¼ or 25%.]
6. Since the whiskers are different lengths but represent the same amount of data, what does that tell you about the pennies in the longer whisker? [Their minting dates are more spread out.]
7. Summarize what the box-and-whisker plot tells you about the minting dates of pennies. [Although the minting dates of pennies may range over more than 30 years, the majority of them have been minted within the last 10 years. In fact, half have been minted since ____ (the median year).]

Application
1. Ask questions like:
   If I get a penny from the store cashier:
   • What is the percent chance that it is older than ____ (the lower quartile year)? [25%]
   • What is the percent chance it is newer than ____ (the median year)? [50%]
2. What are other situations in which you can use this method of gathering data and making predictions?

Extension
Have students study minting dates of different types of coins (nickels, dimes, quarters) and compare them to pennies.

SORTS OF PENNIES

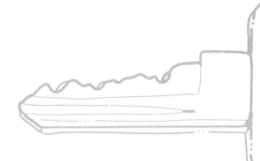

Key Question

In what year was a typical penny currently in circulation minted, and how could you determine that year?

Learning Goals

Students will:

- organize a sample of pennies by minting date;
- determine measures of central tendency: mean, median, mode, and range; and
- construct a line plot and a box-and-whisker plot from the data to study the spread of data.

Sorts of Pennies

In What Year Was a Typical Penny Minted?

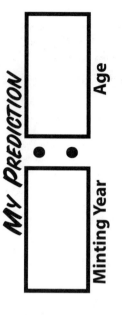

1. Predict the year the typical penny currently in circulation was minted. How old is the typical penny?

2. Sort and stack your pennies into piles that were minted in the same year and put the piles in order from oldest to newest.

3. Transfer the number of pennies in your piles to the line plot with one x for each penny in that year. Leave empty columns for years with no pennies.

Minting Year

Oldest ←————————————————→ Newest

Penny Minting Date Line Plot

What can you say about the typical penny and its minting year or age from looking at the line plot?

SORTS OF PENNIES

IN WHAT YEAR WAS A TYPICAL PENNY MINTED?

Minting Year

Extreme ▶

Newest

Quartile ▶

Median ▶

Quartile ▶

Extreme ▶

Oldest

List the minting date of every penny in your data in sequence from newest to oldest to identify the extremes, quartiles, and median. Then make a box-and-whisker plot of the data.

How does the box-and-whisker plot help you to determine the minting year of a typical penny?

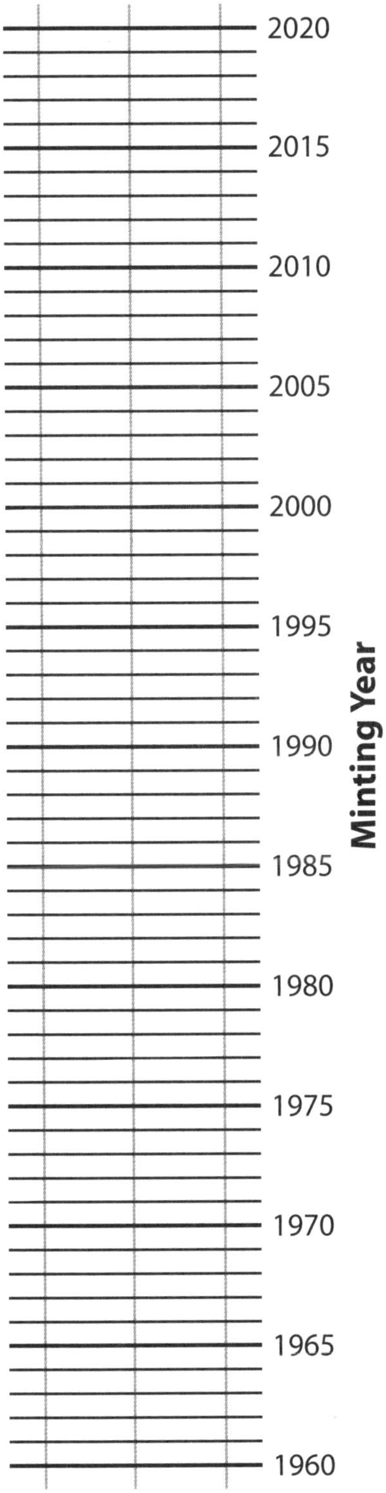

2020

2015

2010

2005

2000

1995

1990

1985

1980

1975

1970

1965

1960

Minting Year

Connecting Learning

Measures of Central Tendency
1. What is the mode of the minting dates of you pennies? ...median? ...range?

2. How do you see it on the line plot?

3. What does it tell you about pennies currently in circulation?

Line Plots
1. What patterns do you see in the heights on the line plot?

2. What does that tell you about these pennies?

3. How does you data compare to the data of other groups?

SORTS OF PENNIES

Connecting Learning

Box-and-Whisker Plot

1. How much of the box-and-whisker plot is made of just the box?

2. What fraction of the pennies does the box represent?

3. Why is half the data filling much less than half the plot?

4. How do the two whiskers of the plot compare in length?

5. What fraction of the pennies does each whisker represent?

6. Since the whiskers are different lengths but represent the same amount of data, what does that tell you about the pennies in the longer whisker?

7. Summarize what the box-and-whisker plot tells you about the minting dates of pennies.

JAMAICAN BOBSLED DATA

Topic
Statistics

Key Questions
How did the times of the Jamaican Bobsled team compare to the Olympic field? Were they competitive?

Learning Goals
Students will:
- use data from the 1988 Calgary Olympics to determine measures of central tendency and spread, and
- construct a box-and-whisker plot to compare the times of competitors.

Guiding Documents
Project 2061 Benchmarks
- *The mean, median, and mode tell different things about the middle of a data set.*
- *Comparison of data from two groups should involve comparing both their middles and the spreads around them.*

NRC Standards
- *Think critically and logically to make the relationships between evidence and explanations.*
- *Mathematics is important in all aspects of scientific inquiry.*

Common Core State Standards for Mathematics *
- *Make sense of problems and persevere in solving them. (MP.1)*
- *Reason abstractly and quantitatively. (MP.2)*
- *Construct viable arguments and critique the reasoning of others. (MP.3)*
- *Use appropriate tools strategically. (MP.5)*
- *Summarize and describe distributions. (6.SP.B)*

Math
Data analysis
    measures of central tendency
        mean, median, mode, range
    data displays
        box-and-whisker plot
    spread

Integrated Processes
Observing
Comparing and contrasting
Collecting and organizing data

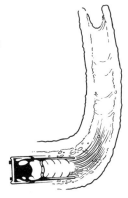

Predicting
Inferring
Interpreting data

Materials
Student page

Background Information
In the 1988 Winter Olympics in Calgary, the Jamaican Bobsled team made its debut. The movie *Cool Runnings* presents the general attitude toward this team from the tropics. Although many looked at the team as a joke, the team came through with final times typical for the range of Olympic results listed below.

1988 Olympic Bobsled Times
Lower Extreme:	55.88 sec.
Lower Quartile:	57.10 sec.
Mode:	56.41 sec.
Mean:	57.6057 sec.
Median:	57.62
Upper Quartile:	58.135
Upper Extreme:	59.67

Jamaican Bobsled Team: 58.04, 59.37

The Olympic Bobsled event is the combined time of four runs down the track. Two runs are made on two different days. To get a final score, all four runs must be completed. The Jamaican team crashed on their fourth and final run so they did not receive an official time. The times are the unofficial times of the first two runs on February 27, 1988.

The finishes of bobsled events are decided by hundredths of a second. The difference of the best and worst times of any event at this level is less than four seconds.

The differences in the ranges of times can be understood by considering the physics of bobsledding. The fastest speed at which a gravity-powered object can move (terminal velocity) is determined by the balance of forces. When the resistive forces of friction and air resistance equal the gravitational force, an object will continue at a constant velocity; it does not accelerate. If two objects, such as bobsleds, have the same resistive forces (four runners and the same frontal area for air resistance), but different gravitational forces (weight of the sled and riders), they will accelerate differently. The sled with the greater mass will overcome the resistive forces more easily, will have a higher terminal velocity, and will accelerate more rapidly.

The rules in the bobsled events are an attempt to maintain equal resistive forces between sleds. In order to win, the team must focus on the critical components of the initial push start, getting in an aerodynamic position, and maintaining a line of least resistance. The difference in Olympic times is a result of these three components.

How would you decide what is the "normal" or "typical" time for a bobsled event? To get at this idea, you might determine the time of a sled in the "middle" of all the times by looking at measures of central tendency: mean (arithmetic average), median (middle of an ordered list), and mode (most often occurring value). These measures will give you only one number. How would you determine the range of very good times? Now you are grappling with the idea of spread. To describe what to expect in a group, you need to have an idea of central tendency and spread for the group.

Spread can be studied by constructing an ordered list of the data and dividing it into four even groups. The value of the position that divides the data into two even groups is the median. The positions that divide the lower quarter and the upper quarter from the middle half are called the lower quartile and upper quartile respectively.

A box-and-whisker plot is one way to display spread. It can be made on graph paper by making a horizontal number line scaled for the range of the data. On a line several squares above the number line, students can place dots to represent the median, the two quartiles, and the slowest and fastest times (extremes). A box can then be drawn around the middle half of the data leaving the quartiles at both ends. A vertical line is drawn in the box at the median position. Lines, or "whiskers," are extended from each end of the box to the extremes.

Box-and-whisker plots can cause some difficulty for students. The students often expect the quarters to be the same size since the data are divided into four even groups. They misinterpret the quarters as the size of each group rather than the range in the value of data for each quarter of the data. Students will often need to consider the meaning of box-and-whisker plots several times before they can clearly interpret them.

Management
1. The data are placed in a sorted list to help students find the median and quartiles. They might find it helpful to circle or mark each measure in a list.
2. If desired, students can use a spreadsheet for this activity. The file of data may be accessed on the accompanying CD.

Procedure
1. Distribute the data sheet to the students and discuss the *Key Question* with the class.
2. Using the data sheet, have the students calculate and record the mean, median, and mode for the 1988 four-man bobsled event.
3. Hold a class discussion as to how the measures of central tendency help answer the *Key Question*.
4. Have students identify and record the median, quartiles, and extremes for the event.
5. Direct the students to construct a box-and-whisker plot for the event.
6. Referring to their box-and-whisker plot, have students draw conclusions about the 1988 Olympic bobsled event and how the Jamaican team would have finished.

Connecting Learning
1. What measure of central tendency does not help you much in making a generalization about the Olympics? [mode, not related to spread]
2. Why is the mean not very useful in drawing conclusions for an Olympic event? [Rank is the only thing that matters in the Olympic decision.]
3. What generalizations can you make for the bobsled events by looking at your box-and-whisker plot?
4. What might explain the differences in the ranges of times? (Refer to *Background Information*.)
5. How competitive was the Jamaican bobsled team in the 1988 Winter Olympics?

Extension
Check statistics from the Winter Olympics subsequent to 1988 to see how the Jamaican two- and four-man bobsled teams did. Official times can be found on the Internet at http://www.todor66.com/olim/index_Winter.html.

Solutions
Mode: 56.41
Mean: 57.57
Median: 57.59
Lower Extreme: 55.88
Lower Quartile: 57.07
Upper Quartile: 58.10
Upper Extreme: 59.07

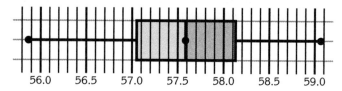

JAMAICAN BOBSLED DATA

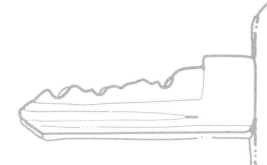

Key Questions

How did the times of the Jamaican Bobsled team compare to the Olympic field? Were they competitive?

Learning Goals

Students will:

- use data from the 1988 Calgary Olympics to determine measures of central tendency and spread, and
- construct a box-and-whisker plot to compare the times of competitors.

JAMAICAN BOBSLED DATA

Official 1988 Olympic Four-Man Bobsled Times

55.88	56.86	57.37	57.82	58.26
56.16	56.93	57.40	57.85	58.28
56.27	57.02	57.41	57.91	58.32
56.33	57.03	57.43	57.91	58.32
56.39	57.07	57.47	57.95	58.35
56.41	57.13	57.50	57.98	58.42
56.41	57.17	57.51	57.98	58.42
56.41	57.18	57.56	58.01	58.49
56.53	57.18	57.58	58.04	58.49
56.66	57.20	57.60	58.04	58.59
56.67	57.20	57.64	58.05	58.65
56.69	57.22	57.66	58.07	58.67
56.70	57.24	57.67	58.10	58.67
56.71	57.25	57.68	58.10	58.80
56.72	57.28	57.69	58.13	58.87
56.72	57.30	57.72	58.14	58.94
56.74	57.31	57.72	58.15	59.02
56.75	57.34	57.75	58.16	59.07

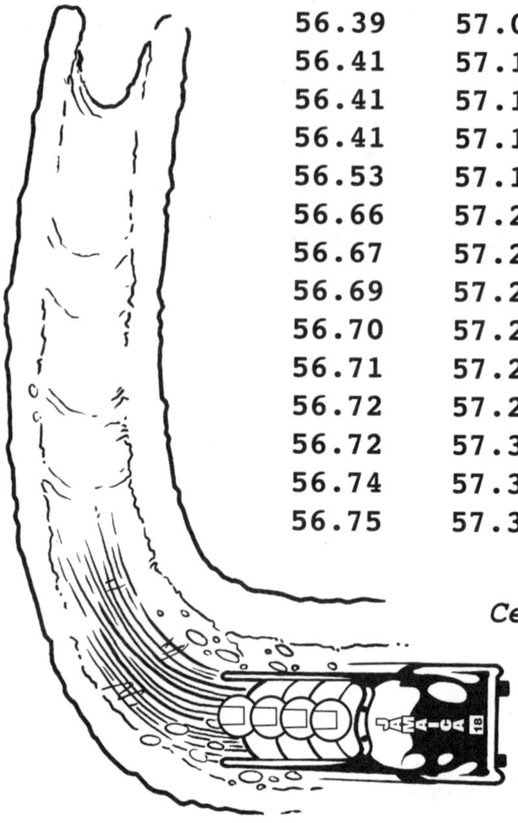

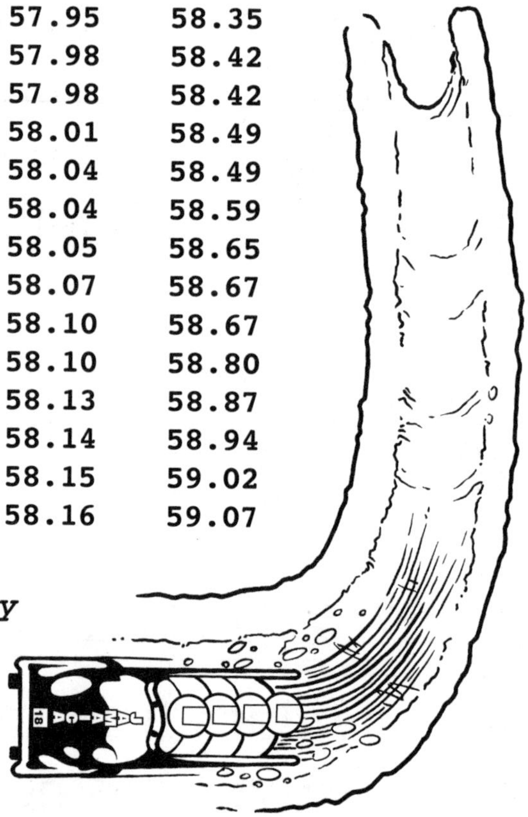

Measures of Central Tendency

Mode

Mean

Median

Lower Extreme | Lower Quartile | Upper Quartile | Upper Extreme

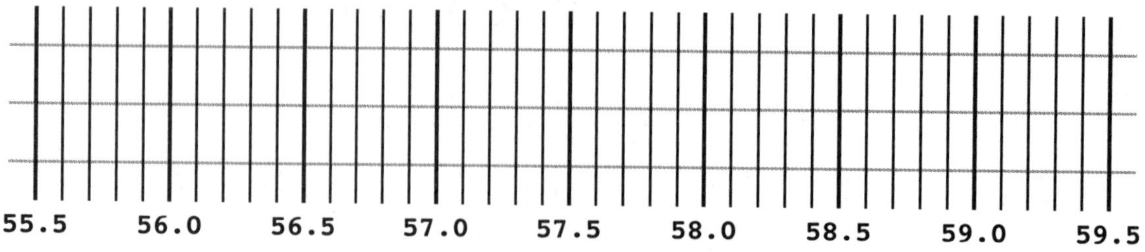

55.5   56.0   56.5   57.0   57.5   58.0   58.5   59.0   59.5

Jamacian Bobsled Team Times: 58.04   59.37

Jamaican Bobsled Data

Connecting Learning

1. What measure of central tendency does not help you much in making a generalization about the Olympics?

2. Why is the mean not very useful in drawing conclusions for an Olympic event?

3. What generalizations can you make for the bobsled events by looking at your box-and-whisker plot?

4. What might explain the differences in the ranges of times?

5. How competitive was the Jamaican bobsled team in the 1988 Winter Olympics?

Topic
Central tendency and spread

Key Questions
1. How tall is the "normal" student in this class?
2. Who is taller in this class, the group of boys or the group of girls?

Learning Goals
Students will:
- measure and sort themselves by height;
- determine the mode, median, quartiles, and range from the ordered arrangement; and
- construct stem-and-leaf plots and box-and-whisker plots from the data.

Guiding Documents
Project 2061 Benchmarks
- *The mean, median, and mode tell different things about the middle of a data set.*
- *Comparison of data from two groups should involve comparing both their middles and the spreads around them.*
- *The larger a well-chosen sample is, the more accurately it is likely to represent the whole, but there are many ways of choosing a sample that can make it unrepresentative of the whole.*
- *What use can be made of a large collection of information depends upon how it is organized. One of the values of computers is that they are able, on command, to reorganize information in a variety of ways, thereby enabling people to make more and better uses of the collection.*

NRC Standards
- *Use appropriate tools and techniques to gather, analyze, and interpret data.*
- *Develop descriptions, explanations, predictions, and models using evidence.*
- *Think critically and logically to make the relationships between evidence and explanations.*
- *Use mathematics in all aspects of scientific inquiry.*

Common Core State Standards for Mathematics *
- *Make sense of problems and persevere in solving them. (MP.1)*
- *Reason abstractly and quantitatively. (MP.2)*
- *Construct viable arguments and critique the reasoning of others. (MP.3)*
- *Model with mathematics. (MP.4)*

- *Use appropriate tools strategically. (MP.5)*
- *Draw informal comparative inferences about populations. (7.SP.B)*

Math
Measurement
Data analysis
    measures of central tendency
        mean, median, mode, range
    data displays
        stem-and-leaf plot, box-and-whisker plot
    spread

Science
Human body
    skeletal proportions
    growth

Technology
Computer applications
    integrated packages
        spreadsheets
        graphing

Integrated Processes
Observing
Collecting and recording data
Organizing data
Interpreting data
Communicating data

Materials
Measuring tape
Masking tape
Sticky notes

Background Information
Statistics are an attempt to condense large bodies of data into digestible pieces. The results often distort the information by not giving the total picture. We read articles about the "average" or "typical" person. But how did the author come up with this number that describes this "average"? Does the number give a true portrayal of the data? If I don't match this "normal" number, how abnormal am I? If students are going to be numerically literate, they are going to need to understand how these numbers are obtained, what they can tell us, and their weaknesses in describing the whole picture.

How would you decide what is the "normal" or "typical" height for a student in your class? To get at this idea, you might determine the height of the student in the "middle" of the class by looking at **measures of central tendency**: mean (arithmetic average), median (middle of an ordered list), and mode (most often occurring value). But if only one value is labeled as "normal," are the rest abnormal? Now you are grappling with the idea of **spread**. To describe what to expect in a group, you need to have an idea of central tendency and distribution for the group.

Stem-and-leaf plots provide a quick way to organize data and get a view of the spread of the data. A piece of data may be broken down into two pieces: the most significant figures, which are called the stems, and the less significant figures, called the leaves. In data about student height, such as 157 cm, we refer to the 15 as the stem, and the 7 as the leaf. By putting the range of stems on one side of a line, each piece of data can be recorded by placing a leaf on the side of the line opposite its corresponding stem. As a large quantity of data is entered, a histogram emerges as the leaves "pile up" along each stem. The difficulty students have in constructing stem-and-leaf plots by themselves is selecting appropriate stem values.

Spread can also be studied by constructing an ordered list of the data and dividing it into four even groups. The value of the position that divides the data into two even groups is the median. The positions that divide the lower quarter and the upper quarter from the middle half are called the lower quartile and upper quartile respectively.

A **box-and-whisker plot** can be made on graph paper by making a horizontal number line scaled for the range of the data. On a line several squares above the number line, students can place dots to represent the median, the two quartiles, and the shortest and tallest students (extremes). A box can then be drawn around the middle half of the data leaving the quartiles at both ends. A

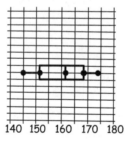

140 150 160 170 180

vertical line is drawn in the box at the median position. Lines, or "whiskers," are extended from each end of the box to the extremes. Making a box-and-whisker plot for several sets of similar data on the same scale provides a comparison. By reorganizing the data and making a box-and-whisker plot of the boys' and the girls' heights, it would be easy to recognize which were typically taller and which had a greater diversity.

Box-and-whisker plots can cause some difficulty for students. The students often expect the quarters to be the same size since the data are divided into four even groups. They misinterpret the quarters as the size of each group rather than the range in the value of data for each quarter of the data. Students will often need to consider the meaning of box-and-whisker plots several times before they can clearly interpret them.

Management

1. Students are to be actively involved in determining the procedures and constructing the displays in this activity. If this is an early experience where the concepts are being developed through more structured instructions, the appropriate tables and graphs are included for modeling. If this experience is for reinforcement and application, students should construct their own tables and graphs using graph paper.

2. This activity has two parts. In the first, students analyze the data of the whole class. This provides an opportunity to develop procedures and instruct and model techniques. The second part has students divide the data into two groups and compare the spread and central tendency. This provides an opportunity for students to work on their own using procedures and constructing displays.

3. To get consistent height measurement, tape a measuring tape (or two, depending on the length of the tape) vertically to a wall with the zero end at floor level. The students should remove their shoes and stand with their backs against the measuring tapes. To obtain a measurement level with the tops of their heads, the edge opposite the binding of a book can be slid down the measuring tape until the bottom of the book rests on the top of the heads.

4. This activity lends itself to computer applications with the use of spreadsheets or databases with graphing capabilities (integrated packages). If this is being used as an application activity, it would be a good time to have students use the computer to record the data and construct the appropriate graphs.

Procedure

Part One

1. Discuss the first *Key Question* with the class.

2. Have the students measure and record their heights on sticky notes.

3. Direct them to arrange themselves in a single file line by height, shortest to tallest. Then have them form the line in a semi-circle.

4. Invite the tallest and shortest students to step out and stand by each other. Discuss the difference in height and relate it to the **range** of the data.

5. While they remain in the line, have students say their heights and ask them to identify where there are the most students of the same height. Relate this to the **mode**.

6. Discuss with the class how they could identify which student is standing in the middle position, the **median**. Methods for doing this might be to count in from both ends or divide the total number of students in half and count in that far from either end.

7. To determine which students form the middle half of the class, have the class find the positions that are a quarter of the class from the shortest and tallest persons. Relate these quarter positions to **quartiles**.

8. Discuss with students that the middle half might describe the typical height better than a single measure of central tendency. Students in the lower quartile might be described as shorter than most, and students in the upper quartile might be described as taller than most students.

9. Have the students place their sticky notes next to each other in order (shortest to tallest) along a line on the board.

10. Discuss how they could determine the range of the data. They should recognize that it is the difference between the first and last height in the ordered string.

11. Invite someone to circle the cluster of sticky notes on the board where there are the most equal heights and identify it as the mode.

12. Have students identify the median and quartiles by marking or circling them on the board. Some students may find the distance between these by dividing the number of sticky notes by four. Others may find it easier to divide the group in half to find the median, and divide the halves in two to find the quartiles. Encourage students to recognize how the data have been split into four equal groups.

13. Invite someone to draw one big box around all the sticky notes that represent the students between the quartiles. Have the students discuss what part of the class this group represents. [the middle half of students by height]

14. Focus attention on those sticky notes outside the box and have students recognize that these represent the shortest and tallest quarters of the class. Inform them that this display provides a representation of a **box-and-whisker plot**.

15. Instruct and model for students how to construct a stem-and-leaf plot and have them transfer the data from the sticky notes to their plots.

16. Using their plots, have students discuss what patterns they see in the spread of heights.

17. Have students identify the median and quartiles using their stem-and-leaf plots. Some students may prefer to transfer the data in the plot into an ordered list (the leaves for each stem are ordered from smallest to largest) because it models their experiences.

18. Direct the students to use the extremes, quartiles, and median to construct a box-and-whisker plot of the data.

140 150 160 170 180

19. Discuss what observations about the data can be made from the box-and-whisker plot.

20. Have students calculate the mean height from the data.

21. Have them summarize how their heights compare to the heights of their classmates.

Part Two

1. Discuss with students whether they think the boys or girls are typically taller in the class (*Key Question 2*).

2. Have the students prepare a two-sided stem-and-leaf plot by putting the girls' leaves on one side of the stems, and the boys' on the other. Discuss how the sides differ and what that tells you about the typical height of each gender.

3. Using the data, have the students determine the extremes, quartiles, and median of each gender and construct a box-and-whisker plot for each.

4. Ask the students to identify any modes and the mean for both groups.

5. Referring to the graphs and statistics, have students summarize how the genders differ in height.

6. Have each student identify where he/she is in each graph and discuss how each compares to the rest of the class.

Connecting Learning

1. What are the strengths or purposes of each of the displays? [stem-and-leaf plot, quick construction of histogram for spread; box-and-whisker plot, comparing spread of data]

2. List all the things you know about a typical student and discuss which displays and measures make each of these things easiest to recognize and show.

3. Find where you are located in each display. How does that position tell you how typical you are?

4. How different do you expect the displays to be for another class at this grade level? How different do you expect the displays to look for a second-grade class?... a class of college students? (Students should recognize a change in the height with age. There should be some discussion on what they expect to happen to spread considering puberty.)

Extension

Measure the heights of children of different ages. Make a box-and-whisker plot for each age and see how the typical height changes from year to year.

Who's Normal?

Key Questions

1. How tall is the "normal" student in this class?
2. Who is taller in this class, the group of boys or the group of girls?

Learning Goals

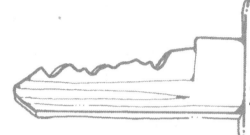

Students will:

- measure and sort themselves by height;
- determine the mode, median, quartiles, and range from the ordered arrangement; and
- construct stem-and-leaf plots and box-and-whisker plots from the data.

Who's Normal?
Part One

Record class height data in the table.

Student Name	Height (cm)
1.	
2.	
3.	
4.	
5.	
6.	
7.	
8.	
9.	
10.	
11.	
12.	
13.	
14.	
15.	
16.	
17.	
18.	
19.	
20.	
21.	
22.	
23.	
24.	
25.	
26.	
27.	
28.	
29.	
30.	
31.	
32.	
Mean	

How tall is the "normal" student in this class?

Organize your data into a stem-and-leaf plot and a box-and-whisker plot.

Stem-and-Leaf Plot

Box-and-Whisker Plot

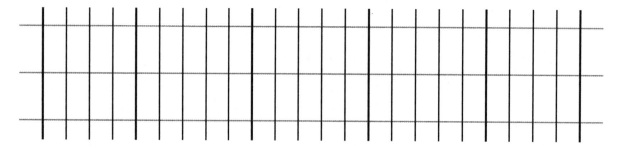

Height (cm)

Who is taller in this class, the group of boys or the group of girls?

Stem-and-Leaf Plot

Box-and-Whisker Plot

Height (cm)

Who's Normal?

Connecting Learning

1. What are the strengths or purposes of each of the displays?

2. List all the things you know about a typical student and discuss which displays and measures makes each of these things easiest to recognize and show.

3. Find where you are located in each display. How does that position tell you how typical you are?

4. How different do you expect the displays to be for another class at this grade level? How different do you expect the displays to look for a second-grade class?... a class of college students?

Color Samples

Topic
Data collection and analysis

Key Question
What quantity of candies and number of each color can you expect in a bag of M&M's®?

Learning Goals
Students will:
- gather the data for eight to 10 bags of M&M's® candies,
- make graphic displays,
- determine measures of central tendency for the samples, and
- summarize their findings to make predictions of other bags.

Guiding Documents
Project 2061 Benchmarks
- *The mean, median, and mode tell different things about the middle of a data set.*
- *Comparison of data from two groups should involve comparing both their middles and the spreads around them.*
- *What use can be made of a large collection of information depends upon how it is organized. One of the values of computers is that they are able, on command, to reorganize information in a variety of ways, thereby enabling people to make more and better uses of the collection.*

NRC Standards
- *Use appropriate tools and techniques to gather, analyze, and interpret data.*
- *Develop descriptions, explanations, predictions, and models using evidence.*
- *Think critically and logically to make the relationships between evidence and explanations.*

Common Core State Standards for Mathematics*
- *Make sense of problems and persevere in solving them. (MP.1)*
- *Reason abstractly and quantitatively. (MP.2)*
- *Construct viable arguments and critique the reasoning of others. (MP.3)*
- *Use appropriate tools strategically. (MP.5)*
- *Use random sampling to draw inferences about a population. (7.SP.A)*

Math
Proportional reasoning
  percent
Data analysis
  measures of central tendency
    mean, median, mode, range
  data displays
    circle graph, bar graph, box-and-whisker plot

Technology
Computer applications
  integrated packages
    spreadsheets
    graphing

Integrated Processes
Observing
Collecting and recording data
Interpreting data
Reporting and communicating data
Generalizing

Materials
Bags of M&M's® candies
Colored pens or pencils, optional

Background Information
  The Mars Candy Company has a set ratio for the colors of M&M's® in a bag. This ratio is maintained for a large batch of bags, but individual bags will vary within a range. By studying a large sample of bags, one can predict the color composition of a typical bag. The company closely monitors the weight of each bag so the quantity of candies is quite regular. There is less variation in the total quantity in each bag than in the individual colors.
  The purpose of this investigation is to introduce students to some concepts and techniques of data analysis in which they will:
- discover the strengths and weaknesses of different displays,
- see the effects of using larger samples of data,
- learn how to find different measures of central tendency, and
- learn to compare the range of different sets of data.

  As students make a circle graph and a bar graph of the data from their group's sample, they can compare the strength of each display. They should recognize the circle graph communicates the ratio within a bag while it is easiest to determine the differences between the colors with the bar graph.
  As students gather the data from each group, they will recognize the differences and will need to

find a way of describing what should be expected for each color. The measures of central tendency—mean, median, and mode—are ways of getting at this "middleness." Students will be quick to observe that all three tend to be nearly the same.

Sorting the different quantities and identifying the high and low (extremes), the middle (median), and quarter divisions (quartiles) provides a better description of the spread of the data. A box-and-whisker plot makes a visual display of this data. Making a box-and-whisker for each color makes it easy to compare them and determine which can be predicted with the most certainty.

Management

1. Allow a minimum of three periods for this activity. One period is needed to gather the data and make the group record. The second and third periods can be used to share group data with the class and complete the statistical analyses.

2. Students should be familiar with finding percents from ratios before beginning this activity.

3. This activity lends itself to computer applications with the use of spreadsheets or databases with graphing capabilities (integrated packages). If this is being used as an application activity, have students use the computer to record the data and construct the appropriate graphs.

Procedure

1. Have students discuss the *Key Question*.

2. Distribute a bag of candy to each group. Inform them not to eat any until all sorting, counting, recording, and verifying is complete.

3. Direct students to sort the candy by color, then to count and record the quantity of each color. Encourage them to verify the count several times before eating the candy.

4. Have students complete the chart by determining the ratios and calculating the percents. Have them construct a circle graph and bar graph from the data. (The circle graph is divided in one percent increments.)

5. Have the students discuss the strengths of the two displays.

6. Allow time for the students to record the class data as groups share their own.

7. Have them compute the class totals and percents, then construct both class graphs.

8. Invite the students to discuss the differences between their group's graphs and the class's graphs.

9. Direct them to determine and record the mean and mode.

10. Have the students use the ordering sheet to order the different quantities of each color and determine and record the median.

11. Ask them to discuss the similarities and meaning of the central tendencies.

12. Using the ordering sheet, direct students to determine the largest and smallest number (extremes) and the quarter dividers (quartiles) for each color.

13. With instruction, have the students construct a box-and-whisker plot for each of the colors and the total.

14. Have students discuss their interpretations of the box-and-whisker plots. Revisit the *Key Question*, encouraging students to apply what is known from the investigation.

Connecting Learning

1. Which graph makes it easiest to answer the question, "How many more red candies are there than green candies?" [bar]

2. Which graph makes it easiest to answer questions about the most and least common colors? [circle]

3. What are the advantages of each of the graphs? [bar, differences between colors; circle, ratio within whole]

4. How are the class graphs similar to and different from your group's graphs?

5. Choose a graph to predict how the colors will be distributed in a new bag. Explain why you chose that graph.

6. What similarities do you see in the mean, mode(s), and median of each color? [nearly the same]

7. How often did a measure of the middle tell what was in the bag? Does one seem to be a better predictor of what is in the bag? Explain.

8. How are the box-and-whisker plots different?

9. Which box-and-whisker plot is the most spread out? What does this tell you about this color?

10. Which box-and-whisker plot is the least spread out? What does this tell you about this color?

11. Predict what will be in a new bag and explain how you came to this decision.

12. You want to show others what you have learned about packages of M&M's®. Choose one display to do it and tell why you chose that display.

Extensions

1. Bring out a large bag of M&M's® and have students use their data to predict the total quantity and number of each color in the bag. After students have predicted what is in the bag, open it and have students check. Ask them to explain any discrepancies and/or accuracies in their predictions.

2. Compare the percent of each color in a bag among different varieties of M&M's® (peanut, dark chocolate, peanut butter, almond, etc.). Are they the same?

Clr Samples

Key Question

What quantity of candies and number of each color can you expect in a bag of M&M's®?

Learning Goals

Students will:

- gather the data for eight to 10 bags of M&M's® candies,
- make graphic displays,
- determine measures of central tendency for the samples, and
- summarize their findings to make predictions of other bags.

Color Samples

Group Data

	Red	Orange	Yellow	Green	Blue	Brown	Total
Count							
Ratio: $\frac{\text{Count}}{\text{Total \#}}$							
Percent of Bag							

Colors in the Bag by Count

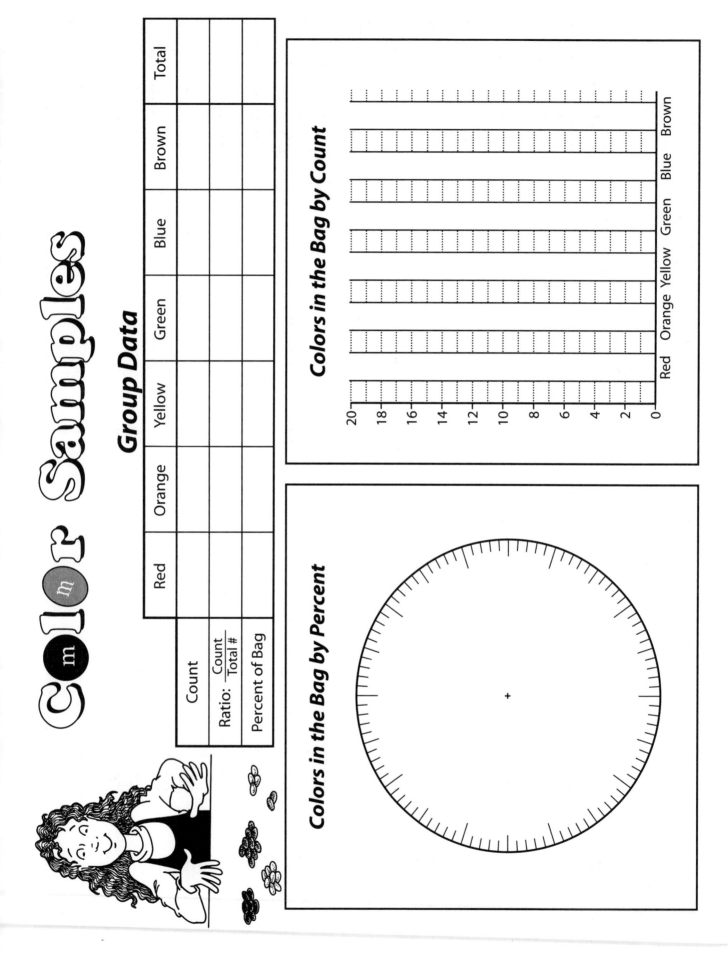

Colors in the Bag by Percent

Color Samples

Class Data

	Red	Orange	Yellow	Green	Blue	Brown	Total
Group 1							
Group 2							
Group 3							
Group 4							
Group 5							
Group 6							
Group 7							
Group 8							
Group 9							
Group 10							
Class Total							
Percent							
Mean							
Mode							
Median							

Color Samples

Class Data Graphs

Colors in the Bag by Count

20						
18						
16						
14						
12						
10						
8						
6						
4						
2						
0	Red	Orange Yellow	Green	Blue	Brown

Colors in the Bag by Percent

Color Samples

1. Order each group's results.
2. Circle the extremes for each color.
3. Determine and mark the quartiles.

Red

Largest

Smallest

Orange

Largest

Smallest

Yellow

Largest

Smallest

Green

Largest

Smallest

Blue

Largest

Smallest

Brown

Largest

Smallest

Total

Largest

Smallest

Box-and-Whisker Plot of Total Candies in a Bag

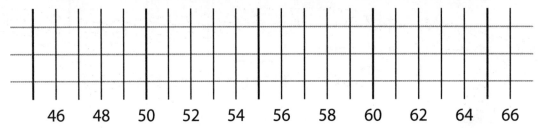

46   48   50   52   54   56   58   60   62   64   66

Box-and-Whisker Plots of Candies by Color

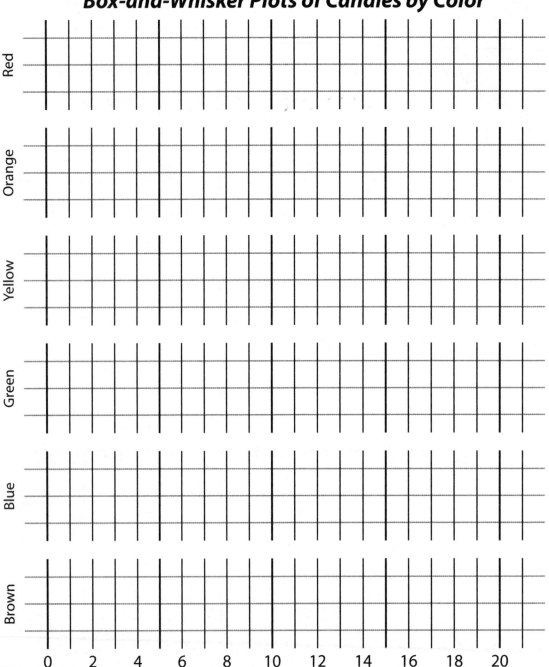

Red

Orange

Yellow

Green

Blue

Brown

0   2   4   6   8   10   12   14   16   18   20

C⬤l⬤r Samples

Connecting Learning

1. Which graph makes it easiest to answer the question, "How many more red candies are there than green candies?"

2. Which graph makes it easiest to answer questions about the most and least common colors?

3. What are the advantages of each of the graphs?

4. How are the class graphs similar to and different from your group's graphs?

Connecting Learning

5. Choose a graph to predict how the colors will be distributed in a new bag. Explain why you chose that graph.

6. What similarities do you see in the mean, mode(s), and median of each color?

7. How often did a measure of the middle tell what was in the bag? Does one seem to be a better predictor of what is in the bag? Explain.

8. How are the box-and-whisker plots different?

Connecting Learning

9. Which box-and-whisker plot is the most spread out? What does this tell you about this color?

10. Which box-and-whisker plot is the least spread out? What does this tell you about this color?

11. Predict what will be in a new bag and explain how you came to this decision.

12. You want to show others what you have learned about packages of M&M's®. Choose one display to do it and tell why you chose that display.

PARKING PLOTS

Topic
Data displays

Key Question
How does a graphic display help you to recognize and communicate the typical range of an attribute (age, state, size, attendance) of all the national parks?

Learning Goals
Students will:
- construct bar graphs, circle graphs, line plots, and box-and-whisker plots;
- choose the appropriate data display for analyzing data and seeing trends; and
- learn which measures of central tendency and spread are most appropriate for describing the data.

Guiding Documents
Project 2061 Benchmarks
- *Organize information in simple tables and graphs and identify relationships they reveal.*
- *Read simple tables and graphs produced by others and describe in words what they show.*
- *Understand writing that incorporates circle charts, bar and line graphs, two-way data tables, diagrams, and symbols.*

Common Core State Standards for Mathematics*
- *Make sense of problems and persevere in solving them. (MP.1)*
- *Reason abstractly and quantitatively. (MP.2)*
- *Construct viable arguments and critique the reasoning of others. (MP.3)*
- *Use appropriate tools strategically. (MP.5)*
- *Summarize and describe distributions. (6.SP.B)*

Math
Data analysis
    measures of central tendency
        mean, median, range
    data displays
        bar graph, circle graph, line plot,
           box-and-whisker plot

Integrated Processes
Observing
Collecting and recording data
Organizing data
Inferring
Interpreting data

Materials
Student pages
Calculators, optional
Colored pencils, optional

Background Information
When we talk about a typical or average national park, we need to find a way to communicate what that means. Measures of central tendency (mean, median, mode) communicate where the middle of the data is. In evenly distributed data, these three measures are the same. But in the data about the national parks, these measures may be far from each other. Sometimes the measures are not appropriate or they misrepresent the truth.

Displays of frequency or spread are often more appropriate for interpreting data than central tendency measures. The bar graph and circle graph are simple displays of frequency data. The line plot, or dot plot, with marks above a number line noting the frequency of the value also gives a sense of spread. The box-and-whisker plot, with the data divided into quarters, allows one to see how the data are spread very quickly. Uneven spread is easily identified by the lack of symmetry in the box and whiskers. In this activity, students will use all of these methods to organize and describe the data.

The box-and-whisker plot is made by sorting the data and then identifying the largest and smallest pieces of data—the extremes. Then, the pieces of data that divide the list into four quarters are determined. The data that divide the quarters are called quartiles, and the middle piece of data is the median. For a box-and-whisker plot, the value of the five pieces of data (minimum, lower quartile, median, upper quartile, and maximum) are marked along a number line. A box surrounds the range of the center half of the data between the two quartiles. Lines (whiskers) are extended from the ends of the box to the extremes.

The state of each park's location can be used to determine the frequency of parks in the state. Some parks are located across several states, so the state with the largest portion is listed. Since no numbers are inherent in the data, the mode as a measure of frequency is the only appropriate measure.

A bar graph or circle graph is the most appropriate display of this non-numeric attribute. It becomes obvious that a few states are disproportionately represented, while most states have no parks at all. Allowing students to consider the information will bring up other observations such as most of the parks (all but eight) are west of the Mississippi River.

	Established	State	Acres	Attendance
Mean	1950.4	N/A	927,803.6	1,101,712.2
Median	1954.5	N/A	242,330	548,004.5
Mode	1980	AK, CA	N/A	N/A

Consider the year each park was established. It seems significant that 1980 saw the greatest number of parks be established—nine. If the age of the parks was being considered, the median of 1945.5 would tell that half our parks were established before that date. Although the mean can be calculated, it would be difficult to establish what it tells us. Using a line plot or a box-and-whisker plot allows us to recognize that the establishment of parks has been a rather steady and continuous process. However, the line plot provides a more meaningful display that provokes questions such as, why were nine parks established in 1980?

With both acres and attendance, there is a great difference between the median and the mean. By looking at these measures along with displays of spread, one gains an understanding of why the extremes skew the mean. In looking at a box-and-whisker plot, it is obvious that there are a few very large or very well attended parks. These might be identified as outliers, or unusually large items of data. Most of the parks have much smaller figures. The box-and-whisker plot displays that half of the parks have fewer than 250,000 acres or 600,000 visitors annually. The huge mean measures are a result of the few outliers of extremely large parks. If one were going to discuss a typical national park, it would be appropriate to exclude the extremes in the mean or look at the median as a more appropriate measure of the middle.

Management

1. The data used for attendance and acreage is from 2007. For the most current data, visit the National Park Service website: http://www.nature.nps.gov/stats/
2. This activity is designed as an application of statistical measures and displays. Students should be familiar with measures of central tendency and spread and their displays. If they are not, modify the activity as an instruction and modeling experience.
3. Depending on the experience of the students, you may want to provide graph paper and allow the students to choose the display that is most appropriate. Student pages are provided to give some structure and a variety of displays for comparison.

4. The use of a spreadsheet is appropriate for this investigation if it is accessible to the students. The file of data may be accessed on the accompanying CD.
5. To simplify the process of determining the quartiles and median, the parks were ordered by year of establishment. The order rank for acres and attendance is included to assist in determining the spread of this data.

Procedure

1. Distribute the list of national parks and have students discuss how they would describe the typical national park from the data.
2. Have students construct a bar graph and circle graph for the national parks by state.
3. Have students discuss or write a summary about conclusions they can draw from the two graphs.
4. Referring to the years parks were established, have students construct a line plot and box-and-whisker plot for the data.
5. Have the students discuss or write a summary about the different conclusions that arise from each of the graphs.
6. Using the data on acres and attendance, have the students calculate measures of central tendency and spread, and make box-and-whisker plots of the data.
7. Have students discuss how the displays provide insights into the data and what conclusions can be drawn.
8. From their investigation, have the students make a summary description of a typical national park.

Connecting Learning

1. Why are the bar graph and circle graphs the appropriate graphs to display the states in which the parks are located? [not numerical, only frequency]
2. What generalizations can you make about the states that have national parks? [States are disproportionately represented; most states have no park; most parks are in the western part of the country.]
3. What are the advantages of the circle graph and bar graph? [circle: gives relative size, more global; bar: distinct numerical difference evident]
4. How do the graphs that show when the parks were established help you decide how old a typical national park is? [median line, box shows range of half the parks, marks on the line plot are relatively evenly spaced, but there have been times of none or many new parks]
5. Which graph do you think makes it easier for you to see how the years are distributed? [choices will vary, personal preference]
6. What observations do you notice in the box-and-whisker plot about acres? ...attendance? [The box and low whisker are very small; the upper quarter of the data is stretched very long.]

7. What do these observations tell you about the parks? [Most parks are lower than 1,000,000 in attendance, but there are a few that are many times larger.]
8. Which measure, mean or median, best describes a typical park? [Median. The median lies in the center or concentration of data. In both cases, the mean is greater than over 3/4 of the parks.]
9. What observation about the graphs helps you understand what caused the mean not to be a very good description of a typical park? [The few extremely large parks skewed the mean.]

Extensions
1. Have students see if there is a correlation between size and attendance by making a scatterplot of acres versus attendance.
2. Have students go to the National Park Service website (http://www.nature.nps.gov/stats/) and find monthly attendance records at each of the most highly attended parks and compare line graphs of attendance by month.

Solutions
National Parks by State

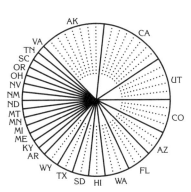

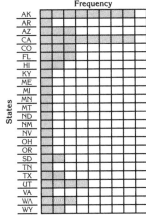

National Parks by Acres and Attendance

	Acres	Attendance
Mean	927,803.6	1,101,712.2
Minimum	5550	847
Lower Quartile	73,563	284,862.5
Median	242,330	548,004.5
Upper Quartile	795,514.5	1,299,565.5
Maximum	8,323,148	9,372,253

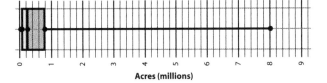

Acres (millions)

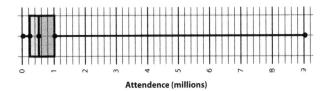

Attendence (millions)

National Parks by Year

Mean	1950.4
Minimum	1872
Lower Quartile	1919
Median	1954.5
Upper Quartile	1980
Maximum	2004

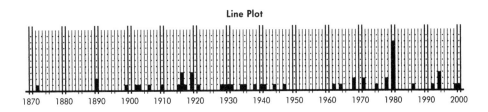

Line Plot

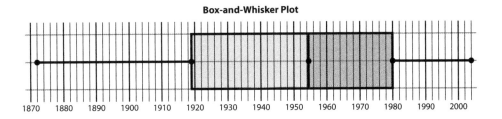

Box-and-Whisker Plot

PARKING PLOTS

Key Question

How does a graphic display help you to recognize and communicate the typical range of an attribute (age, state, size, attendance) of all the national parks?

Learning Goals

Students will:

- construct bar graphs, circle graphs, line plots, and box-and-whisker plots;
- choose the appropriate data display for analyzing data and seeing trends; and
- learn which measures of central tendency and spread are most appropriate for describing the data.

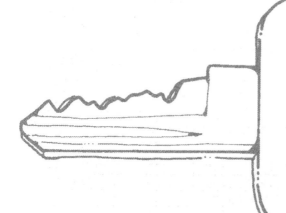

Hawaii Volcanoes National Park

National Parks Data

National Park	Established	State	Size Rank	Acres	Attendence Rank	Attendence
Yellowstone	1872	WY	8	2,219,790	7	2,758,526
Sequoia	1890	CA	22	402,510	23	870,327
Yosemite	1890	CA	16	761,266	4	3,368,731
Mount Rainier	1899	WA	29	235,625	16	1,301,103
Crater Lake	1902	OR	33	183,224	33	457,373
Wind Cave	1903	SD	52	28,295	26	650,357
Mesa Verde	1906	CO	46	52,122	31	513,409
Glacier	1910	MT	13	1,013,572	37	380,114
Rocky Mountain	1915	CO	25	265,723	5	3,139,685
Lassen Volcanic	1916	CA	37	106,372	38	376,695
Hawaii Volcanoes	1916	HI	31	209,695	15	1,343,286
Haleakala	1916	HI	51	29,824	14	1,410,974
Denali	1917	AK	3	4,740,912	40	360,191
Zion	1919	UT	35	146,592	10	2,217,779
Acadia	1919	ME	47	47,633	9	2,516,551
Grand Canyon	1919	AZ	11	1,217,403	2	4,104,809
Hot Springs	1921	AR	54	5549	17	1,296,786
Bryce Canyon	1928	UT	49	35,835	19	1,068,619
Grand Teton	1929	WY	24	309,994	8	2,535,108
Carlsbad Caverns	1930	NM	48	46,766	34	455,621
Isle Royale	1931	MI	18	571,790	51	19,431
Great Smoky Mountains	1934	TN	19	521,621	1	9,197,697
Shenandoah	1935	VA	32	198,081	13	1,498,561
Olympic	1938	WA	14	922,651	3	3,416,069
Kings Canyon	1940	CA	21	461,901	28	541,787
Mammoth Cave	1941	KY	45	52,830	11	1,883,580
Big Bend	1944	TX	15	801,163	41	328,927
Everglades	1947	FL	10	1,508,571	20	1,049,851
Petrified Forest	1962	AZ	38	93,533	27	583,904
Canyonlands	1964	UT	23	337,598	39	368,592
North Cascades	1968	WA	20	504,781	50	27,739
Redwood	1968	CA	36	110,232	36	388,352
Capitol Reef	1971	UT	28	241,904	29	527,760
Arches	1971	UT	42	76,518	24	754,026
Guadalupe Mountains	1972	TX	40	86,416	44	222,307
Voyageurs	1975	MN	30	218,200	43	243,374
Theodore Roosevelt	1978	ND	43	70,447	35	446,609
Badlands	1978	SD	27	242,756	22	955,469
Kobuk Valley	1980	AK	9	1,750,737	54	4217
Lake Clark	1980	AK	7	2,619,733	53	4397
Gates Of The Arctic	1980	AK	2	7,523,898	52	4505
Wrangell–St. Elias	1980	AK	1	8,323,618	49	28,643
Katmai	1980	AK	4	3,674,529	48	67,038
Kenai Fjords	1980	AK	17	669,983	42	262,353
Biscayne	1980	FL	34	172,924	32	489,943
Channel Islands	1980	CA	26	249,354	30	520,428
Glacier Bay	1980	AK	6	3,224,840	12	1,680,614
Great Basin	1986	NV	41	77,180	46	81,712
Dry Tortugas	1992	FL	44	64,701	47	79,186
Saguaro	1994	AZ	39	91,443	25	725,874
Death Valley	1994	CA	5	3,367,628	21	1,014,636
Joshua Tree	1994	CA	12	1,022,703	18	1,280,917
Black Canyon of the Gunnison	1999	CO	53	27,705	45	181,018
Cuyahoga Valley	2000	OH	50	32,860	6	3,123,353

PARKING PLOTS

How many national parks does a typical state have?

Use the national park data to make a bar graph and circle graph showing the number of parks in each state. Label each graph appropriately.

Hawaii Volcanoes National Park

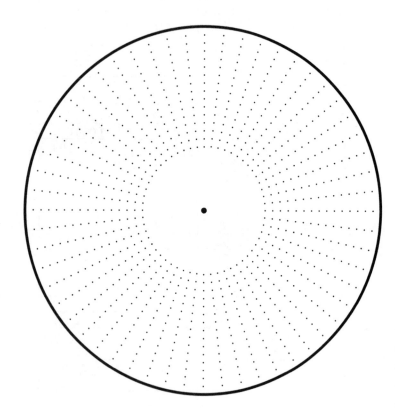

Frequency

By referring to your graphs, what conclusions can you make about the number of national parks in a typical state?

States

PARKING PLOTS: National Parks by Year

In what year was the typical national park established?

Use the national parks data to complete the chart, giving measures of central tendency and spread. Then, make line and box-and-whisker plots of the data.

Mean	
Minimum	
Lower Quartile	
Median	
Upper Quartile	
Maximum	

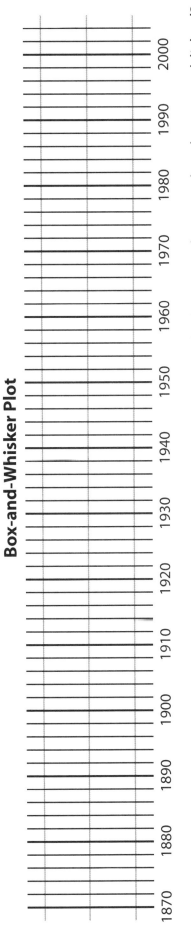

Line Plot

1870 1880 1890 1900 1910 1920 1930 1940 1950 1960 1970 1980 1990 2000

Box-and-Whisker Plot

1870 1880 1890 1900 1910 1920 1930 1940 1950 1960 1970 1980 1990 2000

By referring to your plots, what conclusions can you make about the year in which a typical national park was established?

PARKING PLOTS: National Parks by Acres and Attendance

What is the size of and attendance at a typical national park?

Use the national parks data to complete the chart about measures of central tendency and spread. Then, make box-and-whisker plots of the data.

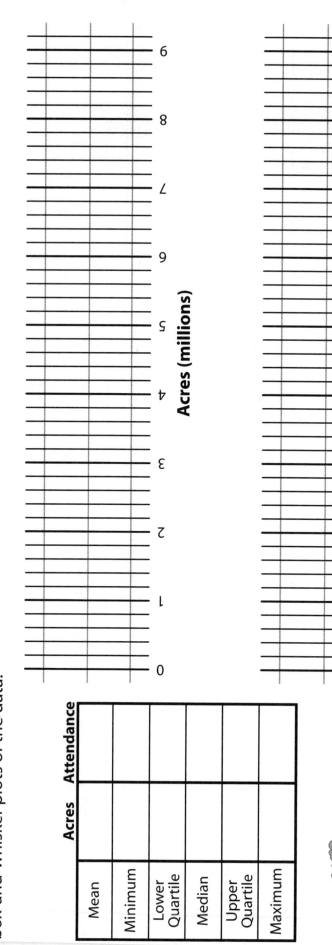

Acres (millions)

6 8 7 6 5 4 3 2 1 0

Attendence (millions)

6 8 7 6 5 4 3 2 1 0

	Acres	Attendance
Mean		
Minimum		
Lower Quartile		
Median		
Upper Quartile		
Maximum		

By referring to your plots, what conclusions can you make about the size of and attendance at a typical national park?

PARKING PLOTS

Connecting Learning

1. Why are the bar graph and circle graphs the appropriate graphs to display the states in which the parks are located?

2. What generalizations can you make about the states that have national parks?

3. What are the advantages of the circle graph and bar graph?

4. How do the graphs that show when the parks were established help you decide how old a typical national park is?

5. Which graph do you think makes it easier for you to see how the years are distributed?

PARKING PLOTS

Connecting Learning

6. What observations do you notice in the box-and-whisker plot about acres? ...attendance?

7. What do these observations tell you about the parks?

8. Which measure, mean or median, best describes a typical park?

9. What observation about the graphs helps you understand what caused the mean not to be a very good description of a typical park?

Drops on a Penny, Revisited

Topic
Central tendency, spread, and correlation

Key Question
How many drops of water can you fit on a penny before it spills over?

Learning Goals
Students will:
- gather data by seeing how many drops of water they can fit on a penny;
- determine range and measures of central tendency—mean, median, and mode; and
- construct stem-and-leaf plots and box-and-whisker plots from the data to study the spread of the data.

Guiding Documents
Project 2061 Benchmarks
- *The mean, median, and mode tell different things about the middle of a data set.*
- *Comparison of data from two groups should involve comparing both their middles and the spreads around them.*

NRC Standards
- *Use appropriate tools and techniques to gather, analyze, and interpret data.*
- *Develop descriptions, explanations, predictions, and models using evidence.*
- *Think critically and logically to make the relationships between evidence and explanations*
- *Use mathematics in all aspects of scientific inquiry.*

Common Core State Standards for Mathematics*
- *Make sense of problems and persevere in solving them. (MP.1)*
- *Reason abstractly and quantitatively. (MP.2)*
- *Construct viable arguments and critique the reasoning of others. (MP.3)*
- *Use appropriate tools strategically. (MP.5)*
- *Draw informal comparative inferences about populations. (7.SP.B)*

Math
Data analysis
measures of central tendency
mean, median, mode, range
data displays
stem-and-leaf plot, box-and-whisker plot

Science
Physical science
adhesion/cohesion
surface tension

Integrated Processes
Observing
Collecting and recording data
Interpreting data
Communicating data

Materials
Pennies
Eyedroppers
Water
Paper towels
Sticky notes

Background Information
Common statistical questions are "What is normal?" "What is typical?" "What can be expected?" All these questions are asking for a generalization about the range and spread of the outcome. Measures of central tendency, such as a mean (average) or median (middle by rank), are numeric ways to express this idea.

The mean, for most students, is a mechanical abstraction with an outcome of little meaning. Initial work with this concept can be done with several stacks of blocks of differing heights. Students can discuss strategies for leveling the heights of the stacks. They might take a block from a tall stack and add the block to a shorter stack. They might choose to put all the blocks together and then restack them evenly into the same amount of even stacks. Experiences like this will help students conceptualize average.

Ordering objects physically and numerically provides a bridge for the understanding of median. The resulting outcome has meaning that can be applied to a situation.

The median brings a partial answer to the question of normal or typical, but because of its exactness, it would seem that very few things are "normal." Splitting a set of data into quarters by rank provides a broader definition of typical. The first and third quarter divisions (quartiles) mark off the middle half of the data. These two middle quarters surround the median and might be a better way to describe normal. The first and last quarters define the lower and upper range of the data. Those ranges may not be extreme or unusual, but they do not describe the common middle of the group.

Graphic displays of the data are visual representations that often make the central tendency and patterns of spread more easily recognized. A *box-and-whisker plot* is made by marking the smallest piece of data, the largest piece of data (these two pieces of data are called the extremes), the two quarter divisions (quartiles), and medians on a number line. A box is drawn between the two quarter divisions to represent the middle half. A line is drawn vertically through the box to indicate the median. A "whisker" line is drawn horizontally from each end of the box to the extremes to show the complete range of the data. The relative size of the box and the extending whiskers give a sense of the range of the data.

A *stem-and-leaf plot* uses numerical data in a display similar to a bar graph. In a piece of numerical data, the more significant units are referred to as the *stems*; the less significant units are referred to as the *leaves*. In the number 35, the 3 would be the stem and the 5 would be the leaf. The range of stems found in the data are listed in a column. Then, each leaf of the data is placed alongside the appropriate stem. Each leaf occupies the same amount of space so that frequency comparisons can be made of the lengths of the rows of leaves. Stems with more leaves in their rows have more data occurring in that range. The differing lengths of rows of leaves makes it very clear how the data are distributed and where most of the data are centered. A sorted stem-and-leaf plot is one in which the leaves for each stem are ordered from smallest to largest.

Water has a tendency to act as if it has a very thin net stretched over its surface. This tendency is called surface tension. Surface tension is caused by the polar nature of the water molecule. The polar charges are caused by electron sharing of the atoms of the molecule. This sharing causes areas of the molecules to become positively and negatively charged. The charged areas are called poles. Opposites charges attract causing the pole on one molecule to be attracted and "cling" to an oppositely charged pole of another molecule. This clinging, or adhesion, of molecules forms the "net" called surface tension.

As water piles up on a penny, it creates a dome-like shape. It will even bulge over the edge of the penny. The "net" of surface tension keeps the water from spilling off. Surface tension will support this bulging dome until the water piles up so high that the force of gravity on the water becomes greater than the strength of the net and it is ripped open as the polar bonds are broken.

Several variables affect the amount of water that can be placed on a penny. One set of variables involves the penny itself: its age, the side used, its condition, etc. A second set of variables includes how the experimenter interfaces with the situation: the technique used for dispensing the drops of water, the height the eyedropper is held above the penny, hand oils or other substances on the penny that affect the adhesive qualities of the water to the penny. (Material in the water, like soap, affects the ability of water molecules to cling to each other decreasing the strength of surface tension.)

Management

1. This activity is written as an introductory experience to the study of central tendency and spread at the middle school level. If it is being used as a reinforcement activity, a more open-ended approach should be used. Although student record sheets are provided, it is suggested that students use graph paper and construct their own record sheets.

2. To make comparisons of data, students will need to have consistent size drops. To encourage this, have students make the smallest drops they can that will freely fall from the dropper. They should place the dropper as close to the water on the penny as possible without disturbing the "dome" of water but still allowing the water to form a distinct drop.

3. Partners work well for this activity. If a larger sample is desired, the students may work individually.

Procedure

1. Have students consider the *Key Question* and make estimates.

2. Direct the students to use the eyedroppers and count how many drops of water they can place on the head side of a penny before it spills over. Discuss with the students the need for consistent size drops for comparison. (See *Management 2*.)

3. Have students record their results on sticky notes.

4. Direct the students to physically order themselves from those with the most drops to those with the least drops in a single file line.

5. As a class, have them determine where to divide the data in half (median), where to divide the greatest quarter of drops and the fewest quarter of drops (quartiles).

6. Draw a horizontal line on the board. Have the students place their sticky notes in order by amount on that line. Mark the median and the quartiles.

7. Have students record the data from the board onto the record sheet and mark the median and quartiles on their list.

8. Using the data, direct the students to record the median, quartiles, and extremes, and determine the mean and mode(s).

9. Discuss with students the similarity and differences of the different types of central tendency.

10. Instruct and model drawing a box-and-whisker plot. Have the students draw their own.

11. Discuss how the box-and-whisker plot compares to their ordering and what new insight can be gained from this display.
12. Explain how to record data on a stem-and-leaf plot. Have the class construct a stem-and-leaf plot as students share the number of drops. Direct the students to then make a sorted stem-and-leaf plot and mark the median, quartiles, and mean.
13. Discuss how the stem-and-leaf plot helps to show the typical number of drops on a penny.
14. Have students return to the original question and summarize how the displays' measures of central tendency help them to answer the question.
15. Direct them to follow a similar procedure to find the answer to "How many drops will fit on the tail side of a penny?"

Connecting Learning
1. Where do you find the fewest drops on each display? ... middle? ... quarter divisions (quartiles)? ... most (extreme)?
2. Which display helps you see the approximate number of drops most pennies hold? [stem-and-leaf plot]
3. Which display helps you to see how spread out the number of drops are? [box-and-whisker plot]
4. Find where the data for your penny is located in each display and discuss how that position tells you how typical it is.
5. Describe how many drops a typical penny will hold. How did you come to this conclusion?
6. Does the side of the penny affect how many drops it will hold? Explain how you used the displays and measures of central tendency to make your conclusion.

Extension
Have students consider other variables (old and new, water and soapy water, clean and dirty pennies) that affect the amount of drops that fit on a penny and gather data to answer their questions.

Key Question

How many drops of water can you fit on a penny before it spills over?

Learning Goals

Students will:

- gather data by seeing how many drops of water they can fit on a penny;
- determine range and measures of central tendency—mean, median, and mode; and
- construct stem-and-leaf plots and box-and-whisker plots from the data to study the spread of the data.

Most

Most

Heads

Tails

Drops on a Penny, Revisited

LOOK OUT BELOW!

THIS IS GONNA HURT!

	Most		Most
Upper Quartile			Upper Quartile
Median			Median
Lower Quartile			Lower Quartile
Least			Least
Mean			Mean
Mode			Mode

Least

Least

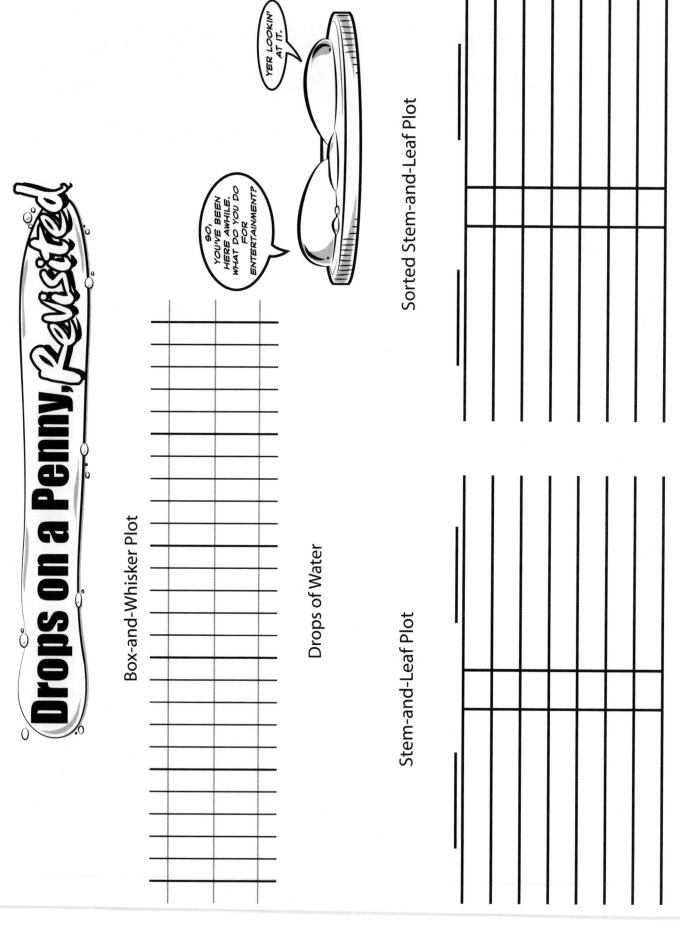

Drops on a Penny, *Revisited*

Box-and-Whisker Plot

Drops of Water

Sorted Stem-and-Leaf Plot

Stem-and-Leaf Plot

Connecting Learning

1. Where do you find the fewest drops on each display? ... middle? ... quarter divisions (quartiles)? ... most (extreme)?

2. Which display helps you see the approximate number of drops most pennies hold?

3. Which display helps you to see how spread out the number of drops are?

4. Find where the data for your penny is located in each display and discuss how that position tells you how typical it is.

113

Connecting Learning

5. Describe how many drops a typical penny will hold. How did you come to this conclusion?

6. Does the side of the penny affect how many drops it will hold? Explain how you used the displays and measures of central tendency to make your conclusion.

LOOK OUT BELOW!

THIS IS GONNA HURT!

Head Size to Hat Size

Topic
Conversion rates

Key Question
How could you tell a person what hat size to buy if you know his or her head size in inches or centimeters?

Learning Goals
Students will:
- gather data about head and hat sizes,
- organize the data in a scatterplot to determine the correlation of hat size to head size, and
- develop ratios to determine the conversion rate from head size to hat size.

Guiding Documents
Project 2061 Benchmarks
- *The expression a/b can mean different things: a parts of size 1/b each, a divided by b, or a compared to b.*
- *Graphs can show a variety of possible relationships between two variables. As one variable increases uniformly, the other may do one of the following: always keep the same proportion to the first, increase or decrease steadily, increase or decrease faster and faster, get closer and closer to some limiting value, reach some intermediate maximum or minimum, alternately increase and decrease indefinitely, increase and decrease in steps, or do something different from any of these.*

Common Core State Standards for Mathematics*
- *Make sense of problems and persevere in solving them. (MP.1)*
- *Reason abstractly and quantitatively. (MP.2)*
- *Construct viable arguments and critique the reasoning of others. (MP.3)*
- *Model with mathematics. (MP.4)*
- *Look for and make use of structure. (MP.7)*
- *Investigate patterns of association in bivariate data. (8.SP.D)*

Math
Data analysis
    correlation
    data displays
        scatterplot
Ratios and proportions
    rates
Linear functions
Measurement

Integrated Processes
Observing
Collecting and recording data
Organizing data
Comparing and contrasting
Predicting
Inferring

Materials
Inch or centimeter measuring tapes
Hat size measuring tape, included
Calculators
Student page (see *Management 2*)

Background Information
Hat sizes in the United States are recorded in eighths of inches between 6⅛ and 7¾. A person's hat size is equal to the diameter of his or her head, if it were a sphere. The circumference of the head (measured just above the eyebrows and across the knob in the back of the skull) is pi (approximately 3.14) times bigger than the hat size. This relationship can be recorded as an equation: Circumference = π · Hat Size or $C = \pi H$. If measurements of a number of people's head circumferences and hat sizes are plotted on a coordinate graph, a line emerges showing the proportional nature of the head size and hat size. To determine a person's hat size, a person's head circumference is measured and divided by pi (≈ 3.14).

If circumference is measured in inches, pi will become evident because hat sizes were developed from our customary system of measurement (inches). If centimeters are used, two ratios combine. The conversion rate of 2.54 cm/inch combines with pi (3.14) to give a factor of 8 ($2.54 \cdot 3.14 = 7.9756 \approx 8$). Using centimeters, students should find that the hat size is about one-eighth of the head's circumference.

When students make a scatterplot of the data, they often find that points do not form a straight line, but rather make a narrow band. This is a result of the error in measurements by students. If they extend a line through the band formed by the data, they will find it passes through the origin. This identifies the proportional nature of the relationship of hat size to head circumference. This proportion is confirmed by the consistent ratio formed by the data.

Management

1. Before class, make hat size measuring tapes and centimeter or inch measuring tapes as required. Copy the measuring tapes and cut them into strips. Use the tabs to glue the strips into measuring tapes. If the tapes are laminated, they can be used for extended periods of time.
2. Before doing the activity with the class, determine if the head circumferences will be measured in inches or centimeters. Copy the appropriate graph page for students.

Procedure

1. Discuss with the class the *Key Question* and how they might determine the relationship of head circumference to hat size.
2. Have students work with partners to measure the circumferences of their heads and their corresponding hat sizes.
3. Distribute the appropriate student page, have students record their own data on the chart, and then have each member of the class share his or her data with the class.
4. Using the class data, have the students make a graph and recognize the linear nature of the data. They should extend the line formed by the data to confirm that the line extends through the origin.
5. Using calculators, have students calculate and record each student's decimal ratio of hat size to head circumference.
6. After students have considered what the graph and the similar ratios tell them about hat size to head size, have them make an equation that models the proportion.

Connecting Learning

1. What pattern do you see in the data points on the graph? [They form a narrow band.]
2. What does the narrow band tell you about hat size and head circumference? [strong positive correlation]
3. If we extend the line formed by the band, noting that it passes through the origin, what does this tell us about the data? [proportional]
4. What do you notice about the ratios of hat size to head circumference? [all nearly the same]
5. Since the ratios are all similar, what is the average ratio and what does it tell you about hat sizes and head circumferences? [about 1/3, or $1/\pi \approx 0.318$, with inches, about 1/8 or 0.125 for centimeters; what fraction of head circumference the hat size is, the conversion rate]
6. What would be an equation that tells someone how to convert head circumference into hat size? [0.318 · head circumference (in) = hat size or 0.125 · head circumference (cm) = hat size]
7. How might you use this outside of school? [sizing apparel and sports equipment, forensic medicine, archeology]
8. What other relationships might there be in skeletal measurements? [foot length to shoe size, height to head circumference, height to foot, legs to total height]
9. What are you wondering now?

Head Size to Hat Size

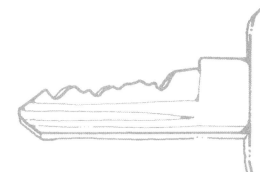

Key Question
How could you tell a person what hat size to buy if you know his or her head size in inches or centimeters?

Learning Goals

Students will:

- gather data about head and hat sizes,
- organize the data in a scatterplot to determine the correlation of hat size to head size, and
- develop ratios to determine the conversion rate from head size to hat size.

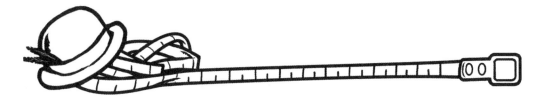

Head Size to Hat Size

1. With the help of a partner, measure and record your head circumference and hat size.
2. Plot the data on the graph and extend the pattern through the whole graph.
3. Calculate the decimal ratio of hat size to head circumference for each student.
4. Explain what the line on the graph and similar ratios tell you about hat size.

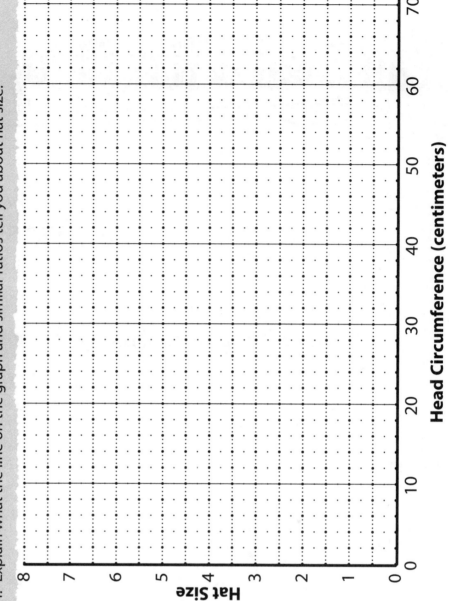

Hat Size (vertical axis: 0–8)

Head Circumference (centimeters) (horizontal axis: 0–70)

Write an equation that tells how to find hat size if the head's circumference is known.

Head Circum. (cm)	Hat Size	Ratio: Hat / Head

Head Size to Hat Size

1. With the help of a partner, measure and record your head circumference and hat size.

2. Plot the data on the graph and extend the pattern through the whole graph.

3. Calculate the decimal ratio of hat size to head circumference for each student.

4. Explain what the line on the graph and similar ratios tell you about hat size.

Head Circum. (Inches)	Hat Size	Ratio: Hat/Head

Hat Size (vertical axis: 0 to 8)

Head Circumference (inches) (horizontal axis: 0, 2, 4, 6, 8, 10, 12, 14, 16, 18, 20, 22, 24, 26)

Write an equation that tells how to find their hat size if the head's circumference is known.

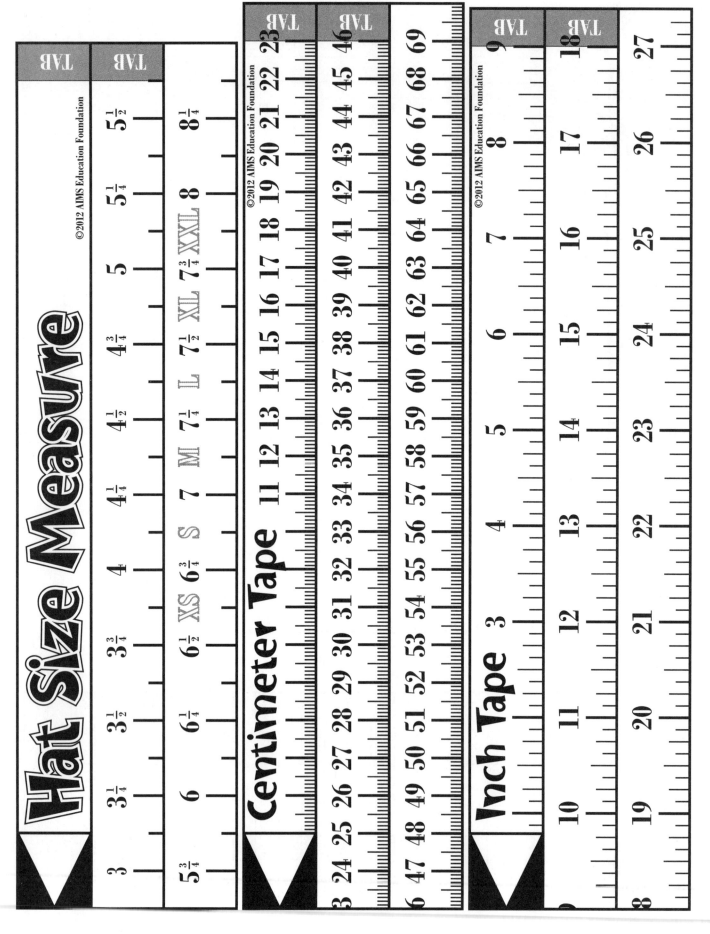

Hat Size Measure

©2012 AIMS Education Foundation

| 3 | 3¼ | 3½ | 3¾ | 4 | 4¼ | 4½ | 4¾ | 5 | 5¼ | 5½ |

| 5¾ | 6 | 6¼ | 6½ | 7 | 7¼ | 7½ | 7¾ | 8 | 8¼ |
| | | | XS | S | M | L | XL | XXL | |

Centimeter Tape 11 12 13 14 15 16 17 18 19 20 21 22 23

©2012 AIMS Education Foundation

24 25 26 27 28 29 30 31 32 33 34 35 36 37 38 39 40 41 42 43 44 45 46

47 48 49 50 51 52 53 54 55 56 57 58 59 60 61 62 63 64 65 66 67 68 69

Inch Tape 3 4 5 6 7 8 9

©2012 AIMS Education Foundation

10 11 12 13 14 15 16 17 18

19 20 21 22 23 24 25 26 27

Head Size to Hat Size

Connecting Learning

1. What pattern do you see in the data points on the graph?

2. What does the narrow band tell you about hat size and head circumference?

3. If we extend the line formed by the band, noting that it passes through the origin, what does this tell us about the data?

4. What do you notice about the ratios of hat size to head circumference?

Head Size to Hat Size

5. Since the ratios are all similar, what is the average ratio and what does it tell you about hat sizes and head circumferences?

6. What would be an equation that tells someone how to convert head circumference into hat size?

7. How might you use this outside of school?

8. What other relationships might there be in skeletal measurements?

9. What are you wondering now?

Ball Bounce

Topic
Data collection and analysis

Key Question
How is the bounce of a ball related to the height from which the ball is dropped?

Learning Goals
Students will:
- drop balls from measured heights and measure the height of the bounce,
- make a scatterplot of the data to identify the correlation, and
- make an equation to model the relationship of bounce height to drop height.

Guiding Documents
Project 2061 Benchmarks
- *Energy appears in different forms. Heat energy is in the disorderly motion of molecules and in radiation; chemical energy is in the arrangement of atoms; mechanical energy is in moving bodies or in elastically distorted shapes; and electrical energy is in the attraction or repulsion between charges.*
- *The expression a/b can mean different things: a parts of size 1/b each, a divided by b, or a compared to b.*
- *Graphs can show a variety of possible relationships between two variables. As one variable increases uniformly, the other may do one of the following: always keep the same proportion to the first, increase or decrease steadily, increase or decrease faster and faster, get closer and closer to some limiting value, reach some intermediate maximum or minimum, alternately increase and decrease indefinitely, increase and decrease in steps, or do something different from any of these.*
- *The graphic display of numbers may help to show patterns such as trends, varying rates of change, gaps, or clusters. Such patterns sometimes can be used to make predictions about the phenomena being graphed.*
- *Use, interpret, and compare numbers in several equivalent forms such as integers, fractions, decimals, and percents.*

Common Core State Standards for Mathematics *
- *Make sense of problems and persevere in solving them. (MP.1)*
- *Reason abstractly and quantitatively. (MP.2)*
- *Construct viable arguments and critique the reasoning of others. (MP.3)*
- *Model with mathematics. (MP.4)*
- *Use appropriate tools strategically. (MP.5)*
- *Attend to precision. (MP.6)*
- *Look for and make use of structure. (MP.7)*
- *Investigate patterns of association in bivariate data. (8.SP.D)*

Math
Measurement
Using rational numbers
    decimals
    percents
    ratios
Data analysis
    measures of central tendency
        mean, median
    data displays
        scatterplot
    correlation
Writing formulae

Science
Physical science
    mechanical energy

Integrated Processes
Observing
Collecting and recording data
Interpreting data
Identifying and controlling variables
Generalizing

Materials
Balls, one per group (see *Management*)
Meter sticks, one per group

Background Information
Mechanical energy, both potential and kinetic, as well as the force of gravity come into play when a ball is dropped and bounces off a surface. The potential energy is converted to kinetic energy as the ball is pulled to the ground by gravity. As the ball hits the floor, that energy is used to deform the ball and the surface it hits. The characteristics of the ball and the

surface determine how much of the energy will be returned to the ball. If either is easily deformed, the energy will be used in deformation. For example, a clay ball makes a splat, or a ball can get stuck in the mud. If the surface and ball are resilient to the impact, most of the energy stays with the ball allowing it to convert that kinetic energy of motion into height as it overcomes gravity on its bounce. The height of the bounce is a measure of the amount of energy remaining with the ball after impact.

The efficiency of a ball and its bouncing surface at conserving energy can be measured by comparing the bounce height to the height from which the ball was dropped. This efficiency remains relatively constant over short drops. As the height of the drop increases, variables of drag and material integrity come into play.

Small variations in bounce height can be expected due to many variables. By doing a number of trials and finding a typical (average) bounce height, a reasonable quantity can be gained. The mean or median both provide methods of coming to the average bounce height. By calculating both, students will see how the two averages compare. The results should be similar.

By determining this average bounce height for a variety of drop heights, a pattern emerges. This pattern is most easily seen graphically. When coordinate pairs of drop height and bounce height are plotted, a linear pattern develops. This linear pattern shows the proportional nature of the relationship.

Numerically, if the average bounce heights are compared in a ratio to the corresponding drop heights, equivalent ratios form. This ratio can often be interpreted better by students in a unit rate, decimal or percent format. If a ball is dropped from 90 cm and bounces 57 cm, it might be easy to explain the ratio with the equivalent of: 3/5, 0.6:1, 0.6, or 60%.

The linear nature of the pattern and its proportional nature means the relationship can be notated symbolically in the form $y = ax$ where (a) is the ratio of bounce height to drop (D) height, (y) is the bounce (B) height, and (x) is the drop height. The equation can be written in a number of equivalent forms: $B = 3/5D = 0.6D = 60\%$ of D. (Note that fractions are rarely used in formulae.)

Management

1. Be sure to clarify the procedures for dropping the balls and reading the bounce heights. The bottom of the ball should be aligned with the drop height on the meter stick. The bounce height should be read from the bottom of the ball. It requires some practice for students to be able to read the height of the ball's bounce. To observe the bounce more accurately, the students should hold a piece of paper or cardboard at eye level in front of the meter stick. When the ball bounces, move the paper to the highest point the bottom of the ball reaches. By sighting over the paper, the student can more easily read the measurement on the meter stick. A "hands and knees" observing position is a must—the observer must be at eye level with the ball at its highest point of the bounce.

2. Another variable to consider is the surface on which the balls bounce. If comparison is to be made among the groups of students, this variable needs to be controlled. It is advisable to have all students use a hard floor as their initial bouncing surface so comparisons are more consistent.

3. Student groups of three or four work well. One can hold the meter stick, one can drop the ball, one can observe the bounce, and one can record the data.

4. Students may collect, record, and compare data for two or more different types of balls. To use a minimal number of balls, groups can trade types after collecting data. Gather a variety of balls that will bounce: golf balls, tennis balls, super balls, rubber balls, etc.

5. If students gather data from more than one ball, have them use the same graph for all the balls with a different colored line to represent each. This provides an opportunity to see the graphic difference of slope and its relationships to the corresponding ratios.

Procedure

1. Have students discuss the *Key Question* and what procedure they would use to get an answer to it.

2. Have a student in each group hold the meter stick vertically with the zero end on the ground. An alternative is to tape the meter stick to the wall with the zero end on the floor.

3. Tell a second student to hold the ball in front of the meter stick so that the bottom of the ball is at a height of 100 cm.

4. After the ball is released, direct the third student to observe and measure the height of the bottom of the ball at the peak of its bounce. Have the students record the data.

5. Direct them to repeat the procedure for a total of five trials at each of the seven drop heights.

6. Have the students complete the record chart by calculating the median and mean bounce heights for each of the seven drop heights.

7. Using the typical bounce height and drop height, have them plot the coordinates on a graph.

8. Have students recognize and discuss the significance of the linear pattern that is formed from the data.

9. Allow time for students to complete the ratio chart by determining the unit ratio and the decimal and percent equivalents for each drop height.
10. Have students discuss the similarity and meaning of the ratios.
11. Using their understanding of the graph and the ratios, direct the students to write an equation that expresses how to find the height of a ball's bounce if the drop height is known.
12. Have students predict how high their ball will bounce if dropped from a given height greater than a meter. Then have them drop the ball from the height to verify the accuracy of their prediction.

Connecting Learning
1. Did any ball bounce as high or higher than the point from which it was dropped? Explain. [No. A ball will never bounce back up to the point from which it was released.]
2. What are some reasons that might explain why a ball cannot bounce back up to the point from which it was released? [Some energy is dissipated in air friction, impact with the floor, and the ball's deformation. If a ball could actually bounce higher than its drop, it could theoretically bounce higher and higher until it bounced into outer space.]
3. What similarities do you notice about the trials at each height? [The bounce height is almost the same.]
4. What do you notice about the mean and median bounce height of each drop height? [The mean and median are very close to the same.]
5. What pattern do you see in the points you plotted on the coordinate graph? [They come close to forming a line.]
6. How can you use this pattern to predict how high your ball will bounce if dropped from 120 cm? ... from 75 cm? [extrapolate, interpolate]

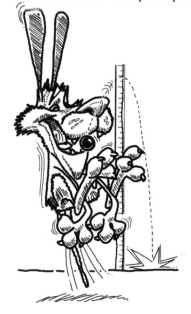

7. A drop height of 50 cm is half the distance of a 100 cm drop. How does the bounce of a 50 cm drop compare to a bounce of a 100 cm drop? [about half as high]
8. What do you notice about all the ratios of bounce to 1 cm of drop? [They are nearly the same.]
9. What does this ratio tell you about the bounce of a dropped ball? [how many centimeters it will bounce for each centimeter it is dropped]
10. How are the decimal and percent equivalents similar to the *Bounce: 1 cm Drop* ratio? [They are equal and mean the same thing; but in the decimal, the ratio to the one unit is assumed, and in the percent it is units per 100 so its format looks slightly different.]
11. If you know how many centimeters a ball is going to drop, how would you determine how far it is going to bounce? [Answers will vary. Students may extrapolate or interpolate from the graph, use combinations of drop heights and bounce heights, or use the unit ratio or decimal or percent equivalents to find a solution.]
12. What is an equation that could be used to predict the bounce height of a ball if the height from which it is going to be dropped is known? [Example: $B = (3/5)D = (0.6)D = (60\%)D$]
13. If you used your equation to predict the height of a ball's bounce that was dropped from the Empire State Building, do you think your prediction would be too high, too short, or just right? Explain and justify your choice. [Predictions will most likely be too high. Factors limit the ball's velocity and resiliency. A terminal velocity limits the energy the ball has and the resulting bounce height.]

Extension
Inflate a basketball with different amounts of air pressure. Find the relationship of air pressure to height of bounce.

* © Copyright 2010. National Governors Association Center for Best Practices and Council of Chief State School Officers. All rights reserved.

Key Question

How is the bounce of a ball related to the height from which the ball is dropped?

Learning Goals

Students will:

- drop balls from measured heights and measure the height of the bounce,
- make a scatterplot of the data to identify the correlation, and
- make an equation to model the relationship of bounce height to drop height.

Ball Bounce

Drop Height(cm)

	100	90	80	70	60	50	40
Trial 1							
Trial 2							
Trial 3							
Trial 4							
Trial 5							
Sum							
Mean							
Median							

Height of Bounce (cm)

Use the table below to help you simplify your bounce to drop ratio. Then, convert the unit ratios to their decimal and percent equivalents.

Ratios

Bounce:Drop	:100	:90	:80	:70	:60	:50	:40
Bounce:10 cm Drop	:10	:10	:10	:10	:10	:10	:10
Bounce:1 cm Drop	:1	:1	:1	:1	:1	:1	:1
Decimal Equivalent							
Percent							

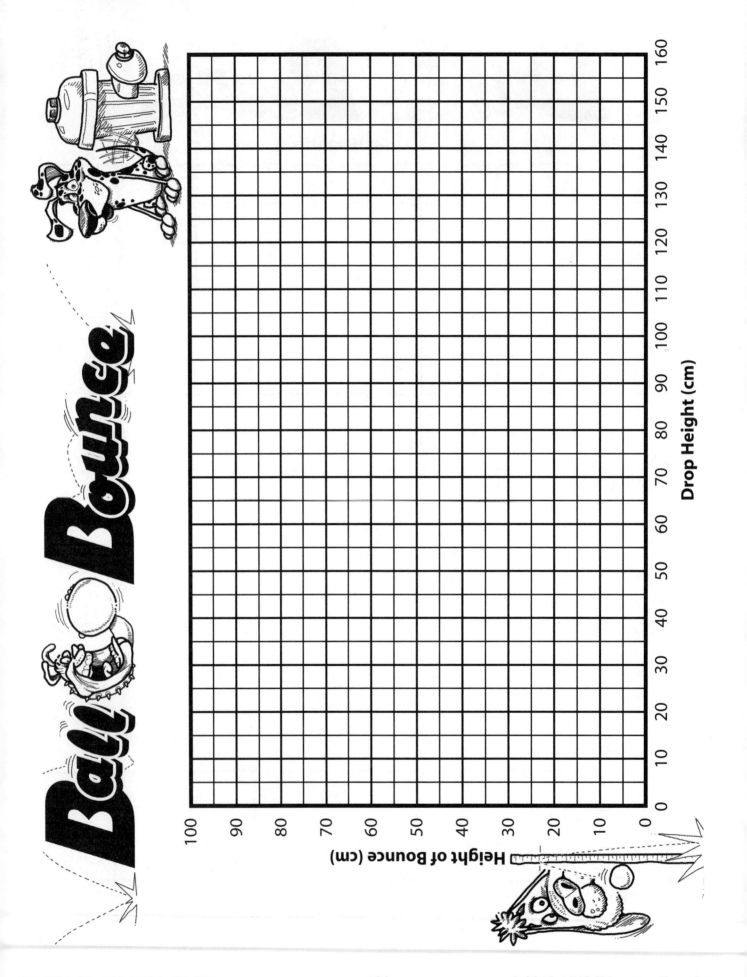

Ball Bounce

Drop Height (cm)

0   10   20   30   40   50   60   70   80   90   100   110   120   130   140   150   160

Height of Bounce (cm)

0   10   20   30   40   50   60   70   80   90   100

Connecting Learning

1. Did any ball bounce as high or higher than the point from which it was dropped? Explain.

2. What are some reasons that might explain why a ball cannot bounce back up to the point from which it was released?

3. What similarities do you notice about the trials at each drop height?

4. What do you notice about the mean and median bounce height of each drop height?

5. What pattern do you see in the points you plotted on the coordinate graph?

6. How can you use this pattern to predict how high your ball will bounce if dropped from 120 cm? … from 75 cm?

7. A drop height of 50 cm is half the distance of a 100 cm drop. How does the bounce of a 50 cm drop compare to a bounce of a 100 cm drop?

8. What do you notice about all the ratios of bounce to 1 cm of drop?

9. What does this ratio tell you about the bounce of a dropped ball?

10. How are the decimal and percent equivalents similar to the *Bounce: 1 cm Drop* ratio?

11. If you know how many centimeters a ball is going to drop, how would you determine how far it is going to bounce?

12. What is an equation that someone could be used to predict the bounce height of a ball if the height from which it is going to be dropped is known?

13. If you used your equation to predict the height of a ball's bounce that was dropped from the Empire State Building, do you think your prediction would be too high, too short, or just right? Explain and justify your choice.

Topic
Central tendency, spread, and correlation

Key Questions
1. How close do you measure up to Leonardo Da Vinci's "ideal" person?
2. How reasonable is it to expect someone to match this ideal image?

Learning Goals
Students will:
- note that Leonardo Da Vinci believed that the ideal person's height was the same as his or her arm span;
- take the required measurements of themselves;
- use stem-and-leaf plots, box-and-whisker plots, scatterplots, line plots, and ratios to gain an understanding of central tendency, spread, and correlation of the data; and
- conclude how "ideal" they think they are.

Guiding Documents
Project 2061 Benchmarks
- *The mean, median, and mode tell different things about the middle of a data set.*
- *Comparison of data from two groups should involve comparing both their middles and the spreads around them.*
- *The larger a well-chosen sample is, the more accurately it is likely to represent the whole, but there are many ways of choosing a sample that can make it unrepresentative of the whole.*
- *Graphs can show a variety of possible relationships between two variables. As one variable increases uniformly, the other may do one of the following: always keep the same proportion to the first, increase or decrease steadily, increase or decrease faster and faster, get closer and closer to some limiting value, reach some intermediate maximum or minimum, alternately increase and decrease indefinitely, increase and decrease in steps, or do something different from any of these.*
- *What use can be made of a large collection of information depends upon how it is organized. One of the values of computers is that they are able, on command, to reorganize information in a variety of ways, thereby enabling people to make more and better uses of the collection.*

NRC Standards
- *Use appropriate tools and techniques to gather, analyze, and interpret data.*
- *Develop descriptions, explanations, predictions, and models using evidence.*
- *Think critically and logically to make the relationships between evidence and explanations.*
- *Use mathematics in all aspects of scientific inquiry.*

*Common Core State Standards for Mathematics**
- *Make sense of problems and persevere in solving them. (MP.1)*
- *Reason abstractly and quantitatively. (MP.2)*
- *Construct viable arguments and critique the reasoning of others. (MP.3)*
- *Model with mathematics. (MP.4)*
- *Use appropriate tools strategically. (MP.5)*
- *Attend to precision. (MP.6)*
- *Look for and make use of structure. (MP.7)*
- *Investigate patterns of association in bivariate data. (8.SP.D)*

Math
Measurement
Proportional reasoning
Data analysis
    measures of central tendency
        mean, median
    data displays
        stem-and-leaf plot, line plot, box-and-whisker plot, scatterplot
    spread
    correlation

Science
Human body
    skeletal proportions

Technology
Computer applications
    integrated packages
        spreadsheets
        graphing

Integrated Processes
Observing
Identifying
Collecting and recording data
Organizing data
Interpreting data
Communicating data
Drawing conclusions

Materials
Metric measuring tapes
Hex-a-link cubes
Masking tape
A book
Graph paper

Background Information
Leonardo Da Vinci was quite right when he generalized that humans' heights are approximately equal to their arm spans. This skeletal proportion is consistent across all age spans of humans. Allowing for a variance of five to 10 percent, we could say that we are all "squares." The following two examples from the world of athletics illustrate one situation that follows Da Vinci's generalization and another that deviates.

- Shawn Bradley of the NBA has a height of 7 feet 6 inches and an arm span of 7 feet 6 inches.
- Michael Gross, the West German Olympic gold medal swimmer of the 1980s, is 6 feet 7 inches tall and has an arm span of 7 feet 4 5/8 inches. (A definite exception to the rule!)

Management
1. Students are to be actively involved in determining the procedures and constructing the displays in this activity. If this is an early experience where the concepts are being developed through more structured instructions, the appropriate tables and graphs are included for modeling. If this experience is for reinforcement and application, students should construct their own tables and graphs with graph paper.
2. This is an extended activity stretching over several days of 45-60 minute periods broken-up as follow:
   - Clarify the question, determine what data are necessary, and how to collect them.
   - Gather and record raw data.
   - Construct and discuss stem-and-leaf plots and box-and-whisker plots.
   - Calculate ratios. Construct and discuss line and box-and-whisker plots of ratios.
   - Construct and discuss scatterplot with mean and median lines.
   - Draw conclusions to findings.

3. To get consistent height measurements, tape a metric measuring tape (or tapes) vertically to a wall with the zero end at floor level. Students being measured should remove their shoes. To get a measurement level with top of the head, the edge of a book can be slid down the measuring tape(s) until the book rests on top of the head. The measurement should be read from the bottom of the book.
4. To get consistent arm span measurements, hang a measuring tape or tapes horizontally along a wall at the shoulder height of the shortest student. A small block (e.g., Hex-a-link cube) can be taped so one face is at the zero end of the measuring tape. The students being measured should stand facing away from the measuring tape, spread their arms, and move until their shoulders are up against the measuring tape and their longest finger is touching the block. While in this position, slide another block along the tape until it touches the longest finger on the other outstretched hand.
5. This activity lends itself to computer applications with the use of spreadsheets or databases with graphing capabilities (integrated packages). If this is being used as an application activity, have students use the computer to record the data and construct the appropriate graphs.
6. If students need more experience with organizing and interpreting data, you may choose to look at height and arm span data separately on different days before moving to the ratio and correlation of the two. (Refer to *Who's Normal?*)

Procedure
Identifying the Question and Procedure
1. Have students look at the Leonardo Da Vinci-type illustration and discuss what proportions are illustrated. Have them determine that one proportion shown is that a person's height should be the same as the arm span.
2. Ask the *Key Questions* and discuss them in light of this proportion. Have students clarify and state in their own words what they are trying to determine.
3. Talk about what information they need to know to get the answer to this question. Elicit the response that they need the heights and arm spans of a sample of individuals.
4. Discuss with the students how they can get the measurements they need. They need to consider how precisely they want measurements taken (to the nearest centimeter is sufficient for this activity), and how they are going to guarantee the measurements are made consistently. (*Management* gives some suggestions.)

Collecting and Organizing Data
1. Have the students discuss what data they need to record in an appropriate table.
2. Direct them to measure and record the data for each student.
3. Have them make a stem-and-leaf plot for the heights and one for the arm spans from the data. (It is effective to make these back-to-back.) When the plots are completed, have students discuss how these displays help answer the question.
4. Urge the students to sort the data in numeric order on the second stem-and-leaf plot. Have them use this new plot to determine the median and quarter divisions of the data. Using this information, have them construct box-and-whisker plots for both the height and arm span data. Ask what new insights or advantages there are in the box-and-whisker plot.
5. Have students calculate the average (mean) for height and arm span. Lead a class discussion about the similarity and difference between the mean and the median.
6. Direct the students to calculate the decimal equivalent of the height-to-arm span ratio for each student. Discuss the meaning of this ratio.
7. Have the students make a line plot of the ratio data and discuss the patterns and their significance.
8. Guide them to determine the median and quarter divisions and construct a box-and-whisker plot of the ratio data. Discuss the similarities and differences of the information conveyed by the line plot and box-and-whisker plot.
9. Instruct students to construct a scatterplot from the height and arm span data. To get the most useful display, students will need to adjust the scale so the data are spread throughout. The same scale should be used on both the height and arm span scales. Have the students discuss what information is conveyed with this display. They should also discuss the meaning of correlation and determine whether there is a correlation of the data, and if so, whether it is positive (slants upward to the right) or negative (slants downward to the right).
10. Have the students find points to construct a median and mean line. (They can get the coordinates for these points by multiplying a given arm span by the median and mean ratios to get the corresponding height.) When several points for the median and mean have been determined, the lines can be constructed and labeled on the scatterplot. Students may also find it useful to plot a line for the "ideal" person. Have them discuss the similarities and differences in the lines and what insights the lines gave them to the meaning of the scatterplot.

Communicating Results
1. Have students summarize all the things they have found out relating to the question.
2. Allow time for them to compare and contrast the displays, discussing which best related different aspects of the answer.
3. Hold a class discussion concerning the limits to their findings. Will the finding be true for all ages and groups?
4. Ask students to discuss how they might go about the data collection differently to get a better answer.
5. Have students list other questions that have arisen from their exploration.

Connecting Learning
1. What are the strengths or purposes of each of the displays? [stem-and-leaf—spread; box-and-whisker plot—spread; line plot—spread; scatterplot—correlation]
2. List all the things you know about a typical student and discuss which displays make each of these things easiest to recognize and show.
3. Find where you are located in each display and discuss how that position tells you how "ideal" you are.
4. How different do you expect the displays to be for another class in this grade? How different do you expect them to look for a second grade class? … a class of college students? (Students should recognize a change in the height and arm spans with age. There should be some discussion on what they expect to happen to the proportional relationship.)
5. How could you gather data differently to get a better data?

Extensions
1. Measure the height and arm spans of different age children and check the consistency of proportion.
2. Make other skeletal measurements and see what proportions there are.

* © Copyright 2010. National Governors Association Center for Best Practices and Council of Chief State School Officers. All rights reserved.

Close to Ideal

Key Questions

1. How close do you measure up to Leonardo Da Vinci's "ideal" person?
2. How reasonable is it to expect someone to match this ideal image?

Learning Goals

Students will:

- note that Leonardo Da Vinci believed that the ideal person's height was the same as his or her arm span;
- take the required measurements of themselves;
- use stem-and-leaf plots, box-and-whisker plots, scatterplots, line plots, and ratios to gain an understanding of central tendency, spread, and correlation of the data; and
- conclude how "ideal" they think they are.

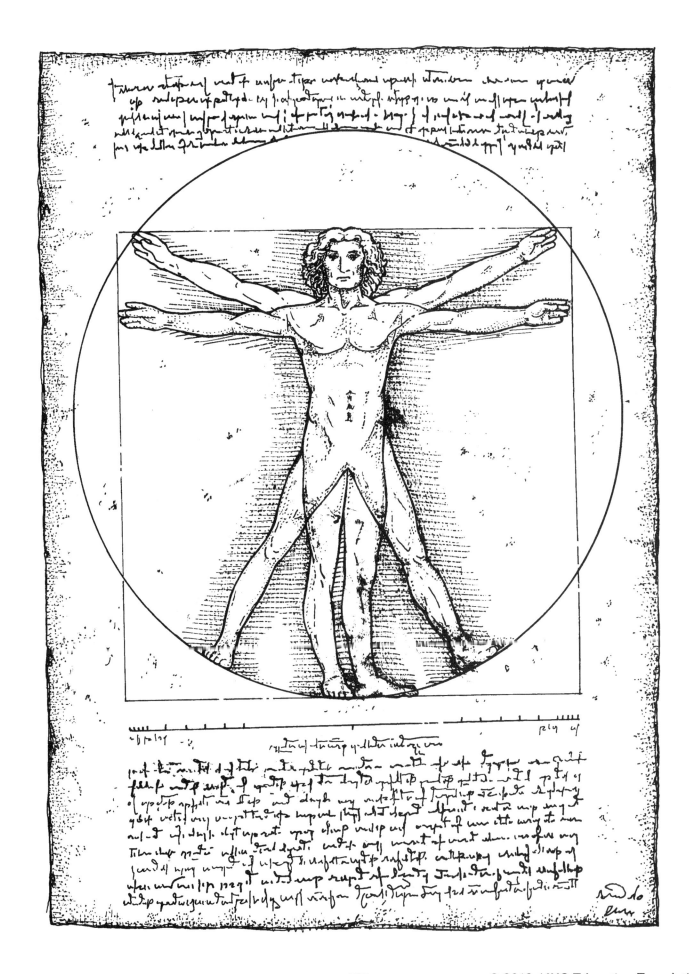

Close to Ideal

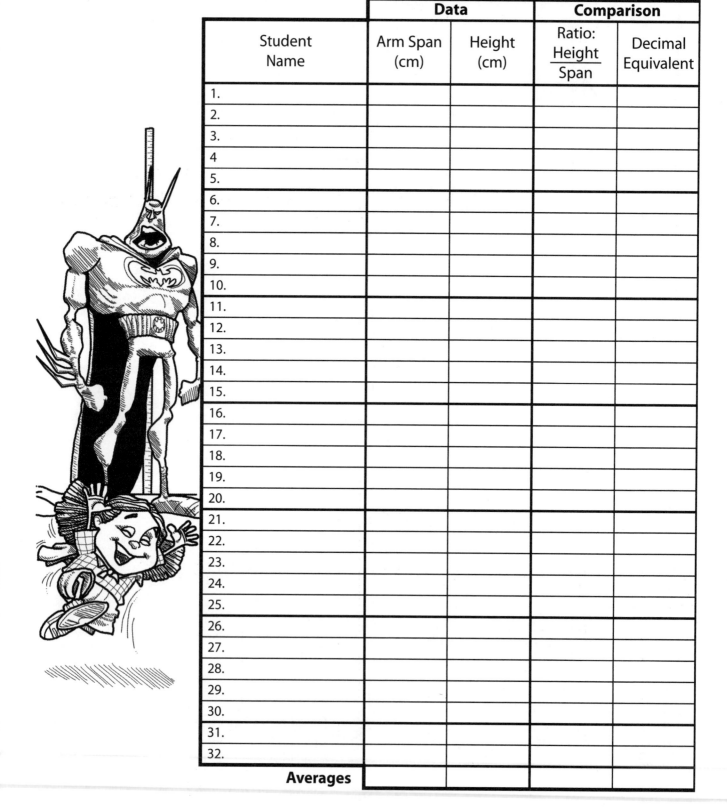

Student Name	Data		Comparison	
	Arm Span (cm)	Height (cm)	Ratio: $\frac{Height}{Span}$	Decimal Equivalent
1.				
2.				
3.				
4				
5.				
6.				
7.				
8.				
9.				
10.				
11.				
12.				
13.				
14.				
15.				
16.				
17.				
18.				
19.				
20.				
21.				
22.				
23.				
24.				
25.				
26.				
27.				
28.				
29.				
30.				
31.				
32.				
Averages				

Height vs. Arm Span Ratio Line Plot

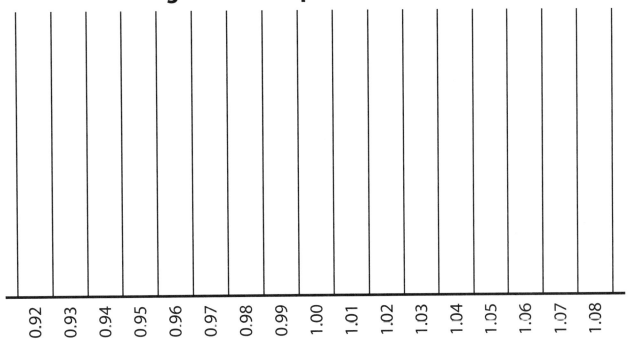

Ratio

0.92 0.93 0.94 0.95 0.96 0.97 0.98 0.99 1.00 1.01 1.02 1.03 1.04 1.05 1.06 1.07 1.08

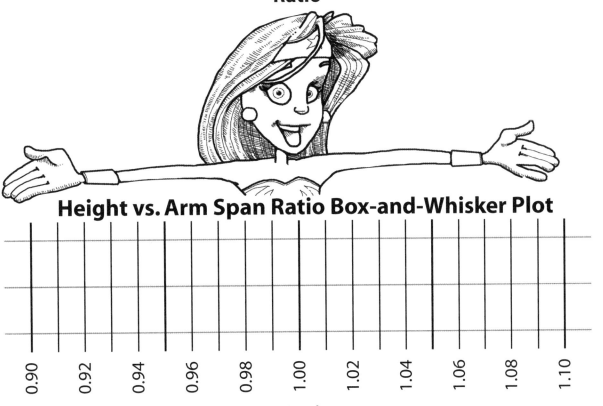

Height vs. Arm Span Ratio Box-and-Whisker Plot

Ratio

0.90 0.92 0.94 0.96 0.98 1.00 1.02 1.04 1.06 1.08 1.10

Close to Ideal

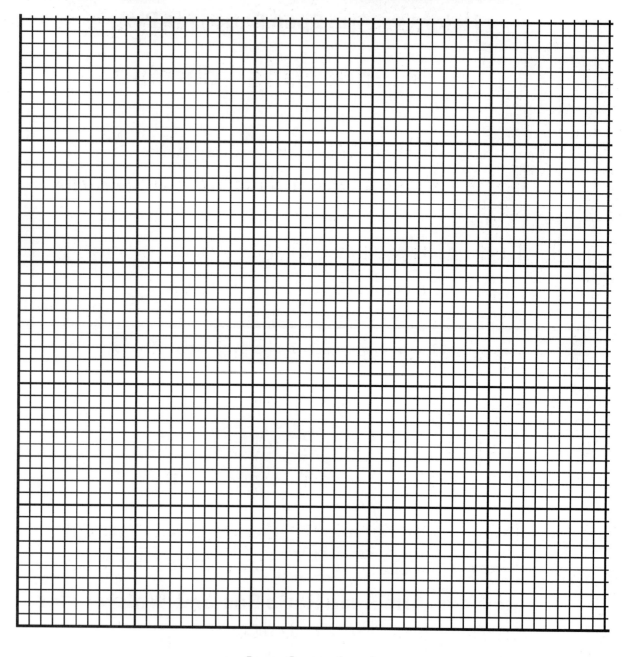

Height (cm)

Arm Span (cm)

Arm Spans	140 cm	150 cm	160 cm
Predicted Heights Using the Mean Ratio			
Predicted Heights Using the Median Ratio			

Mean Ratio:	
Median Ratio:	

Connecting Learning

1. What are the strengths or purposes of each of the displays?

2. List all the things you know about a typical student and discuss which displays make each of these things easiest to recognize and show.

3. Find where you are located in each display and discuss how that position tells you how "ideal" you are.

4. How different do you expect the displays to be for another class in this grade? How different do you expect them to look for a second grade class? ... a class of college students?

5. How could you gather data differently to get a better data?

Hats, Heads, and Heights

Topic
Statistics, correlation

Key Question
If you found the hat of a student, how might you determine that student's height?

Learning Goals
Students will:
- take measurements of class members' heights and head circumferences,
- analyze the data using both a scatterplot and averages, and
- determine if there is a correlation between a person's height and head circumference.

Guiding Documents
Project 2061 Benchmarks
- *The expression a/b can mean different things: a parts of size 1/b each, a divided by b, or a compared to b.*
- *Graphs can show a variety of possible relationships between two variables. As one variable increases uniformly, the other may do one of the following: always keep the same proportion to the first, increase or decrease steadily, increase or decrease faster and faster, get closer and closer to some limiting value, reach some intermediate maximum or minimum, alternately increase and decrease indefinitely, increase and decrease in steps, or do something different from any of these.*
- *The graphic display of numbers may help to show patterns such as trends, varying rates of change, gaps, or clusters. Such patterns sometimes can be used to make predictions about the phenomena being graphed.*

Common Core State Standards for Mathematics*
- *Make sense of problems and persevere in solving them. (MP.1)*
- *Reason abstractly and quantitatively. (MP.2)*
- *Construct viable arguments and critique the reasoning of others. (MP.3)*
- *Use appropriate tools strategically. (MP.5)*
- *Attend to precision. (MP.6)*
- *Look for and make use of structure. (MP.7)*
- *Investigate patterns of association in bivariate data. (8.SP.D)*

Math
Measurement
   linear
Proportional reasoning
Data analysis
   measures of central tendency
      mean
   data displays
      scatterplot
   correlation
Writing equations

Science
Life science
   human body
      skeletal proportions

Integrated Processes
Observing
Collecting and recording data
Interpreting data
Comparing and contrasting
Generalizing

Materials
Metric measuring tapes
Student pages

Background Information
This activity explores one of the many ratios evident in the human body. It also shows the deceptive nature of estimating circumferences.

A mature adult's height approaches three times the head's circumference. The comparison of height to head circumference in early adolescents (ages 10-14) produces a ratio of about 2.8. This ratio decreases with age. Early school age students (ages 5-8) have a ratio of about 2.2. The ratios are more consistent for adults who have completed their periods of growth.

It must be established that there is a correlation between two sets of data before one set can be used to predict the other. Because there is a greater variance in ratios for children, they are not as quick to recognize the correlation found in this activity. In order to help them recognize this correlation and provide them with a way to verify any correlation, they need to be introduced to the scatterplot.

A scatterplot is a statistical tool used to see if two sets of data are correlated. On a coordinate grid, a point is plotted for each corresponding pair of data.

The closer this set of points comes to forming a line, the stronger the correlation is between the two sets of data. The more scattered this set of points becomes, the weaker the correlation.

Height to head circumference is a positive correlation. It is called a positive correlation because as the head size increases, the height also increases. It would be a negative correlation if one set of data changed inversely to the other set. A negative correlation would exist if, as head size got bigger, height decreased. When using a scatterplot, the data's plotted points for a positive correlation will form a cloud of dots that drifts upward as it moves across the graph to the right. The upward drift shows the positive correlation. The "thickness" of the cloud communicates the strength of the correlation. The narrower this band of points, the stronger the correlation.

A line of best fit is a line that come closest to being in the center of the cloud of plotted points. It is appropriate for the middle-school student to estimate and draw this line on a scatterplot. The student can recognize that the slope of this line is close to the average ratio of height-to-head circumference. This line will go up the average ratio amount of units each time the line goes to the right one unit. The equation for this line is: Height = (Average Ratio x Head Circumference). This is the same as the common algebra equation: $y = ax$, where y is the vertical axis, x is the horizontal axis and a is the slope.

The scales on the graph for this activity have been set up to help students recognize the positive correlation of the data.

Management

1. This activity is enhanced if before the activity the teacher gets a hat that is sized to a specific person whose height is known. An adjustable baseball cap works well. This hat is used as an example in introducing the activity. At the end of the activity, it is used again as students predict the height of the hat's owner.
2. If metric measuring tapes are not accessible, string can be wrapped around the head and measured with a meter stick.
3. Make sure all students are taking consistent head measurements. A good suggestion is to measure across the eyebrows, above the ears, and across the bump on the back of the head.
4. Allow two periods for this activity. One period is needed for gathering and sharing the data. (The graphing can be done at home.) The second period is reserved for analyzing and applying the data.
5. This activity works well with students working with partners.

Procedure

1. Introduce the situation and discuss the *Key Question*.
2. Have students predict how they think their heights compare to their head circumferences.
3. Direct students to measure and record their heights and head circumferences to the nearest centimeter.
4. Allow time for them to share their data with the class.
5. Direct students to make a scatterplot from the data.
6. Have students make the comparison of height to head circumference for all the individuals.
7. Ask students to calculate the averages for each type of data.
8. Take time to discuss what patterns they find in the scatterplot and how the average ratio relates to the graph.
9. Have students determine a way to predict the height of a student if the head size is known.
10. Direct students to record the average ratio in the second column on the prediction chart and use it to predict the height for each head size.
11. Have them place an X on the scatterplot to represent each head circumference and predicted height pair.
12. Discuss the significance of the line of best fit formed by the Xs.

Connecting Learning

1. What patterns can you identify on the graph? [Answers will vary. Most notice the points line up in vertical columns. Depending on the data, some will recognize that the data fall in a band.]
2. Why do the points never fall between the vertical lines on the graph? [Measurements were taken in whole centimeters.]
3. If someone told you his head had a circumference of 55 cm, how would you predict his height from the graph? [One can only tell the likely range of height.]
4. Have students trace a line around the outside of the data points and ask them: What happens to the area as it goes across the page to the right? [It goes up.]
5. If the area goes up as the head sizes get bigger, what does that say about the height? [It also goes up.]
6. What does this graph tell us about the comparison of height to head circumference? [As the head gets bigger, a person gets taller. (Tell the students that this is called a correlation.)]
7. How does the narrowness or thickness of the area of plotted data on the graph affect how you can predict the height of a person? [The narrower the band, the more accurate the prediction will be.]

8. What do you notice about the decimal equivalents for all the people? [They are close to the same.]

9. What does the average decimal ratio tell you? [The middle ratio for the class. About how many times taller a person is than his or her head's circumference.]

10. How can you use this ratio to determine a person's height if you know his or her head size? How can you write your explanation as a equation? [To get the height (H) you multiply the ratio (R) by the head circumference (C). H = (R) x (C)]

11. Predict the height of a person of each head circumference and plot an X on the graph to represent the predicted height for that head size. What do you notice about all the Xs? [They form a line that goes through the middle of the data points.]

12. How might a correlation like this help at a crime scene or in an archeological dig?

Extensions
1. Size the hat for a student in kindergarten or first grade. Discuss how the ratio of height to head circumference changes with age.

2. Have students gather data of different aged students to see if the ratio of height to head circumference changes with age. Direct students to make a scatterplot and find an average for each age. Students may find a box-and-whisker plot provides a clearer comparison of all the data.

3. Have students determine the median ratio and the mode ratio and see how they compare to the average (mean) ratio. If there is a difference, does it produce a more accurate prediction?

* © Copyright 2010. National Governors Association Center for Best Practices and Council of Chief State School Officers. All rights reserved.

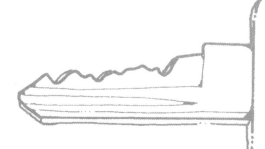 **Key Question**

If you found the hat of a student, how might you determine that student's height?

Learning Goals

Students will:

- take measurements of class members' heights and head circumferences,
- analyze the data using both scatterplot and averages, and
- determine if there is a correlation between a person's height and head circumference.

Class Results

Student Name	Data		Comparison	
	Head Circumference (cm)	Height (cm)	Ratio: $\dfrac{\text{Height}}{\text{Head}}$	Decimal Equivalent
1.				
2.				
3.				
4.				
5.				
6.				
7.				
8.				
9.				
10.				
11.				
12.				
13.				
14.				
15.				
16.				
17.				
18.				
19.				
20.				
21.				
22.				
23.				
24.				
25.				
26.				
27.				
28.				
29.				
30.				
31.				
32.				
Averages				

Hats, Heads, and Heights

Hats, Heads, and Heights

1. Record the head circumferences and heights of all the students in your class on the *Class Results* page.

2. On the graph page, make a scatterplot with a point representing each student.

3. Calculate the decimal equivalent ratio for each student and find the class averages.

4. What patterns do you see in the graph?

5. What does the average decimal equivalent ratio tell about the heights and heads of students in your class?

Head Circum. (cm)	Average Ratio	Predicted Height (cm)
48		
49		
50		
51		
52		
53		
54		
55		
56		
57		
58		
59		
60		
61		
62		

6. How could you use the average ratio to determine the height of someone if you knew his or her head circumference?

7. Record the average ratio in the second column of the chart to the right and use it to determine the predicted height.

8. Put an X on the scatterplot to represent the predicted height of each head circumference.

Hats, Heads, and Heights

If you know the circumference of a person's head, how can you predict his or her height?

What patterns do you see in the graph?

Height (cm)

200
190
180
170
160
150
140
130
120
110
100

48 49 50 51 52 53 54 55 56 57 58 59 60 61 62

Head Circumference (cm)

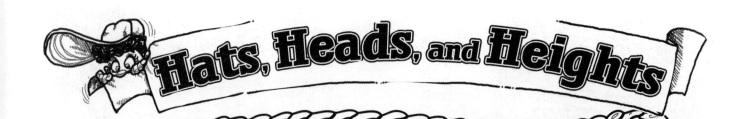

Connecting Learning

1. What patterns can you identify on the graph?

2. Why do the points never fall between the vertical lines on the graph?

3. If someone told you his head had a circumference of 55 cm, how would you predict his height from the graph?

4. What happens to the area as it goes across the page to the right?

5. If the area goes up as the head sizes get bigger, what does that say about the height?

6. What does this graph tell us about the comparison of height to head circumference?

Connecting Learning

7. How does the narrowness or thickness of the area of plotted data on the graph affect how you can predict the height of a person?

8. What do you notice about the decimal equivalents for all the people?

9. What does the average decimal ratio tell you?

10. How can you use this ratio to determine a person's height if you know his or her head size? How can you write your explanation as a equation?

11. Predict the height of a person of each head circumference and plot an X on the graph to represent the predicted height for that head size. What do you notice about all the Xs?

12. How might a correlation like this help at a crime scene or in an archeological dig?

Topic
Probability

Key Question
What is the probability of choosing from a large number of paths the one path that leads to a pot of gold coins?

Learning Goals
Students will:
- select one of 12 paths and follow the selected path to determine if it leads to a pot of gold;
- use such words as certain, equally likely, and impossible to describe the experience;
- compute the theoretical probability of selecting the correct path; and
- perform a probability experiment.

Guiding Documents
Project 2061 Benchmarks
- *Probabilities are ratios and can be expressed as fractions, percentages, or odds.*
- *Events can be described in terms of being more or less likely, impossible, or certain.*

Common Core State Standards for Mathematics*
- *Make sense of problems and persevere in solving them. (MP.1)*
- *Reason abstractly and quantitatively. (MP.2)*
- *Construct viable arguments and critique the reasoning of others. (MP.3)*
- *Investigate chance processes and develop, use, and evaluate probability models. (7.SP.C)*

Math
Data analysis
    probability

Integrated Processes
Predicting
Observing
Recording
Comparing and contrasting

Materials
Colored pencils
Scissors
Transparent tape or glue stick
Student pages

Background Information
This activity helps students identify that the probability of choosing from 12 possible paths the one correct path to a goal lies somewhere *between* impossible (0) and certain (1). The probability of choosing the correct path is therefore represented as a fraction greater than zero but less than one.

If a quarter, dime, and penny are the only coins sitting on the top of a table, it is *impossible* to choose a nickel from the group of coins. If one coin is selected from the group, it is *certain* to be either a quarter, dime, or penny. In mathematical probability, an impossible event is assigned the value of zero. An event certain to happen is assigned the value of one. Intermediate events are greater than zero but less than one.

Simple probability is the first topic in their study of mathematics where students meet two answers to a problem; a theoretical answer and an answer arrived at by experiment. The two answers don't always agree.

The experimental answer is the answer that counts; it's the answer that determines winners and losers. The theoretical answer is useful in predicting results only if the experiment is performed a large number of times. The larger the number of experiments, the closer the results will be to the theoretical answer (this is called *The Law of Large Numbers*).

A *probability experiment* is defined as a procedure that has the same set of possible outcomes every time it is repeated. No single outcome of the experiment is predictable.

The set of all possible outcomes of a probability experiment is defined as the *sample space* of the experiment. An *event* is a subset of a sample space. In this activity, the sample space is the set of path numbers {1, 2, 3, 4, 5, 6, 7, 8, 9, 10, 11, 12}. Picking path number 3 is an event, since {3} is a subset of the sample space.

Theoretical probability is defined as the chance or likelihood that a certain event in a sample space will occur.

In this activity, students choose one of 12 paths in the hopes that the path leads to a cave containing a pot of gold coins. The theoretical probability of picking the one path that leads to the pot of gold coins is one out of 12, or one-twelfth.

Management

1. Make enough copies of the *Inner Maze* pattern page so that every student will have one inner maze pattern. Separate each pattern on the page by cutting along the dashed lines. This will save time distributing the patterns to students.

2. Before beginning this activity, determine the cave that will hold the pot of gold coins. Write the cave number on a piece of paper, fold the paper in half twice and place the paper where all can see it.

Procedure

1. Explain to the students that a pot of gold coins is in one of 12 caves and that a single path leads to each cave. Tell them that the number of the cave containing the pot of gold coins is written on the folded paper in plain sight, for all to see.

2. Distribute scissors and one *Inner Maze* pattern to each student. Instruct the students to carefully cut around the outer edge of the circle.

3. Explain to the students that they are going to perform a probability experiment. Instruct students to choose the numbered path they think leads to the pot of gold and to shade or color that path to the edge of the circle.

4. Distribute one *Outer Maze* page to each student. Tell the students to place the circular piece in the center of the page and to rotate the piece until each of the 12 paths line up correctly.

5. Distribute glue sticks or transparent tape and instruct the students to carefully glue or tape the circular piece in place.

6. Tell the students to continue coloring or shading the path they chose until it reaches one of the 12 caves.

7. Use this chart of the path connections to check that students are tracing the correct path from the path number they chose to the correct cave number.

Path #	leads to	Cave #	Path #	leads to	Cave #
1	→	3	7	→	8
2	→	9	8	→	6
3	→	11	9	→	4
4	→	1	10	→	12
5	→	10	11	→	5
6	→	7	12	→	2

8. Collect the student pages and tape them up around the classroom. Open the folded sheet of paper and announce the winning cave to the class. Move along the display of student pages and mark the winners.

9. Distribute one *What are your chances?* page to each student.

10. Have the students write why it's not impossible for them to choose the correct path and explain why it is not certain they will choose it.

11. Ask the students to shade the rectangle, starting from the *Impossible* end, according to what they believe their chances are of winning.

12. Instruct the students to answer the remaining questions and compute the probability of choosing the winning path.

13. Ask the students to select one of the words printed above the rectangle to describe their chance of winning.

Connecting Learning

1. Why is it not impossible for you to find the pot of gold coins? [There is at least one path to the coins.]

2. Why is not certain you will find the pot of gold coins? [There is more than one path from which to choose.]

3. Explain why you shaded the probability rectangle the way you did.

4. How many of the 12 possible paths lead to the pot of gold coins? [one]

5. How many possible paths are there? [12]

6. What is the theoretical probability of choosing the winning path? [1/12]

7. Describe, in one word, your chance of picking the correct path. [unlikely]

8. How many times would you have to play the game to expect to win once? [12 times]

9. If the teacher had also been allowed to choose a path to the gold coins, would it have been fair? Explain.

10. Count the total number of winners in the class and divide by the number of students that performed the experiment. This is the *experimental* probability. Compare the theoretical probability (1/12) or its decimal equivalent to the experimental probability.

Extension

Ask students in as many other classes as possible to perform the probability experiment. Collect the data and compare the number of successful picks to the total number of experiments. The larger the number of students involved, the closer the ratio will be to 1/12, or its decimal equivalent, 0.083333.

Key Question

What is the probability of choosing from a large number of paths the one path that leads to a pot of gold coins?

Learning Goals

Students will:

- select one of 12 paths and follow the selected path to determine if it leads to a pot of gold;
- use such words as certain, equally likely, and impossible to describe the experience;
- compute the theoretical probability of selecting the correct path; and
- perform a probability experiment.

Inner Maze Patterns

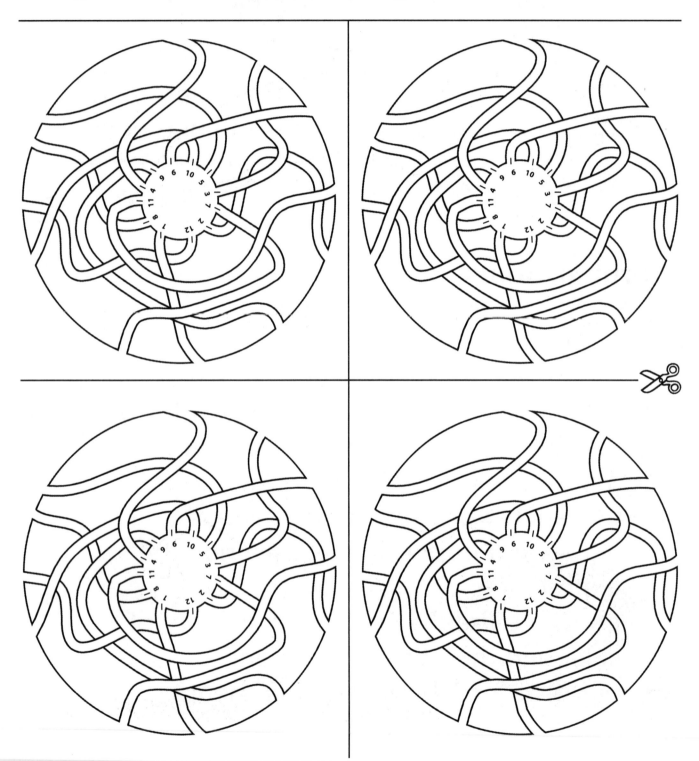

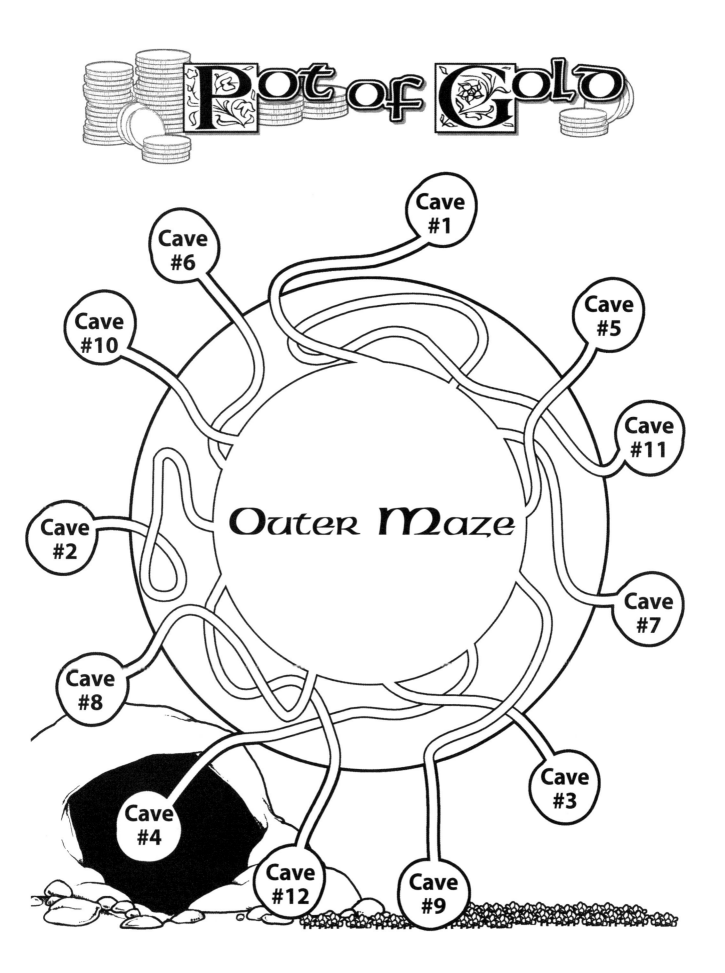

Pot of Gold

Outer Maze

Cave #1
Cave #6
Cave #10
Cave #5
Cave #2
Cave #11
Cave #8
Cave #7
Cave #4
Cave #3
Cave #12
Cave #9

Pot of Gold
What are your chances?

Why is it not impossible for you to choose the path that leads to the pot of gold coins?

Why is it not certain that you will choose the path that leads to the pot of gold coins?

Beginning at the left end of the rectangle below, use a colored pencil to shade the part of the rectangle you think indicates your chance of finding the pot of gold coins.

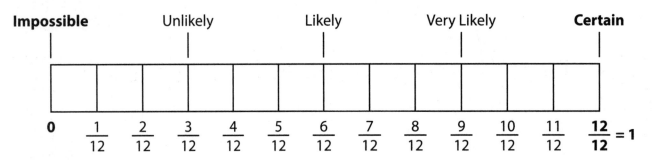

How many posssible paths are there?

How many of the possible paths lead to the pot of gold coins?

Let the probability of choosing the winning path = P. Compute the probability of P.

$$P = \frac{\text{How many winning paths?}}{\text{How many possible paths?}} = \underline{\hspace{2cm}}$$

Use one of the words above the shaded rectangle to describe your chance of picking the winning path.

Connecting Learning

1. Why is it not impossible for you to find the pot of gold coins?

2. Why is not certain you will find the pot of gold coins?

3. Explain why you shaded the probability rectangle the way you did.

4. How many of the 12 possible paths lead to the pot of gold coins?

5. How many possible paths are there?

6. What is the theoretical probability of choosing the winning path?

Connecting Learning

7. Describe, in one word, your chance of picking the correct path.

8. How many times would you have to play the game to expect to win once?

9. If the teacher had also been allowed to choose a path to the gold coins, would it have been fair? Explain.

10. Count the total number of winners in the class and divide by the number of students that performed the experiment. This is the *experimental* probability. Compare the theoretical probability (1/12) or its decimal equivalent to the experimental probability.

FAIR PLAY

Topic
Probability

Key Question
Are the two games you are playing fair?

Learning Goals
Students will:
- play a game that involves making moves on a game board based on the outcome of drawing cubes from a bag;
- keep track of number of wins and ties; and
- determine whether it is likely, unlikely, or equally likely for players to win the race.

Guiding Documents
Project 2061 Benchmark
- *Events can be described in terms of being more or less likely, impossible or certain.*

Common Core State Standards for Mathematics*
- *Make sense of problems and persevere in solving them. (MP.1)*
- *Reason abstractly and quantitatively. (MP.2)*
- *Construct viable arguments and critique the reasoning of others. (MP.3)*
- *Investigate chance processes and develop, use, and evaluate probability models. (7.SP.C)*

Math
Data analysis
    probability
    data collection
Estimation
Problem solving

Integrated Processes
Observing
Collecting and recording data
Comparing and contrasting
Generalizing

Materials
For each pair of students:
    game boards
    bag with cubes (see *Management 2*)
    game pieces

Background Information
Probability is a measure of uncertainty; it is a measure of likelihood of an uncertain event; it is a measure with values between 0 and 1. These are key ideas that we want students to come to understand.

One of the ways that students can express the measure of uncertainty in a situation is by using words such as likely, unlikely, and equally likely, or using their own words that have the same meaning. The bicycle race that is the focus of this activity is driven by a very common kind of probability experiment—drawing cubes (or other objects) from a bag.

For the first challenge race, each rider will be assigned a color. They will then take turns drawing a cube from the bag and moving ahead if they draw the assigned color. Since the bag contains two red and two yellow cubes, the riders have an equal chance of winning.

For the second challenge race, each rider will draw two cubes from the bag. Rider #19 will move one space if the two cubes are the same color, and rider #37 will move one space if the two cubes are different colors. While it might at first appear that the riders in this race are equally likely to win, that is not the case. There are only two ways to draw a pair of cubes from the bag that are the same color; however, there are four ways to draw a pair of cubes of different colors. This can be seen if the individual cubes are identified. For example, the cubes might be labeled r1, r2, y1, and y2. Then, the possible pairs where the cubes are the same color are {r1, r2} and {y1, y2}. The possible pairs where the colors are different are {r1, y1}, {r1, y2}, {r2, y1}, and {r2, y2}. When the second challenge race has been completed, students should have the opportunity to see that in a sense, the race is not fair—the riders is in this race don't have an equal chance of winning.

Management
1. The activity is designed for pairs of students.
2. Each pair needs a paper bag containing four cubes in two colors. If cubes are not available, other objects that come in two colors and are identical in shape can be substituted. The cubes (or objects) need to be labeled so that they are distinguishable. If the objects are yellow and red, then they might be named y1, y2, r1, and r2.
3. The page of bicycle pieces has enough for seven pairs of students. Copy the page onto card stock, cut the pieces apart, and fold them along the dashed lines so that they stand upright.
4. For clarity, you may wish to color (or have students color) the #19 pieces red and the #37 pieces yellow.
5. Make a transparency of the class results page to use when compiling group data.

Procedure

1. Tell the students that they will be playing a board game that involves a bicycle race and that their progress in the race will be determined by drawing cubes (or other objects) out of a bag.

2. Have students get into pairs and distribute the materials for the first challenge race. Have students look at the contents of the bag to see that it contains two yellow and two red cubes.

3. Ask that one student in each group draw a cube from the bag. If the cube drawn is red, that student will be rider #19; if it is yellow, he/she will be #37.

4. Tell students that they will be taking turns drawing a cube from the bag. After drawing a cube and seeing its color, the cube will be returned to the bag. Rider #19 will draw first and will move one space if the cube is red. Then rider #37 will draw a cube and will move one space if the cube is yellow. If rider #19 draws a yellow cube, he/she does not move. Likewise, if rider #37 draws a red cube, he/she does not move.

5. Explain that they will be taking turns until one or both of them reaches the finish space. It's important that the players have equal numbers of turns. If rider #19, who had the first turn, crosses the finish line first, rider #37 must still be given one more turn to assure that they have an equal number of turns. A race is finished when one (or both) of the riders reaches the finish space.

6. Tell the students that they will be riding in six races and that they are to record the winners and ties on their papers.

7. When all of the pairs of riders have completed their six races, ask them to report how their races turned out. Use the transparency of the class results sheet to record the groups' data. Discuss whether or not this race was fair and why.

8. Give each pair the game page for the second challenge race.

9. Tell students that this race is similar to the first one, except they will be drawing two cubes at a time out of the bag, which still contains the same four cubes. Show them that not only are the cubes two different colors, but they are named y1, y2, r1, and r2.

10. Explain the rules for the second race: Rider #19 gets to draw cubes first. If he/she draws two cubes of the same color, he/she moves bicycle #19 one space. The cubes are returned to the bag and rider #37 gets to draw. Rider #37 gets to move his/her bicycle one space if, whenever he/she draws, the colors of the two cubes are different. The cubes are returned to the bag.

11. Allow time for the students to again complete six races and record the winners/ties.

12. Put up the class results transparency and have groups once again share the winners from their games. Discuss whether or not this game was fair and why.

13. Have each pair of students look at the cubes more closely and ask them to make a record of all of the possible pairs that can be drawn. They should come up with {y1, r1}, {y1, r2}, {y1, y2}, {y2, r1}, {y2, r2}, and {r1, r2}. You might want to tell them that these six pairs are the set of *all possible outcomes* as the result of drawing two cubes at a time and then replacing them.

14. Direct students to list the outcomes where rider #19 would move and the ones where rider #37 would move. Students should find that four out of six of the possible outcomes favor rider #37 and only two out of six favor #19.

15. Discuss the fairness of this version of the game.

Connecting Learning

1. In your group, who won the first race most often? Did you ever tie?

2. Based on our class results, did both players have an equal chance of winning?

3. If we had played the game 10 times instead of six, do you think the results would have been different? Why or why not?

4. Was the first race fair? Did both players have an equal chance of winning? Why or why not?

5. Describe the probable outcome of the first race using words like *likely*, *unlikely*, or *equally likely*. [It is equal for either player to win.]

6. In your group, who won the second race most? Did you ever tie? Did #19 ever win?

7. Based on our class results, did both players have an equal chance of winning the second race?

8. Was the second race fair? Why or why not?

9. Describe the probable outcome of the second race using words like *likely*, *unlikely*, or *equally likely*. [It is unlikely that #19 will win the race; it is likely that #37 will win the race; it is not equally likely that either player will win.]

10. How does listing the possible outcomes help you see if a race is fair or not?

11. What would you do to make the second race fair?

* © Copyright 2010. National Governors Association Center for Best Practices and Council of Chief State School Officers. All rights reserved.

FAIR PLAY

Key Question

Are the two games you are playing fair?

Learning Goals

Students will:

- play a game that involves making moves on a game board based on the outcome of drawing cubes from a bag;
- keep track of number of wins and ties; and
- determine whether it is likely, unlikely, or equally likely for players to win the race.

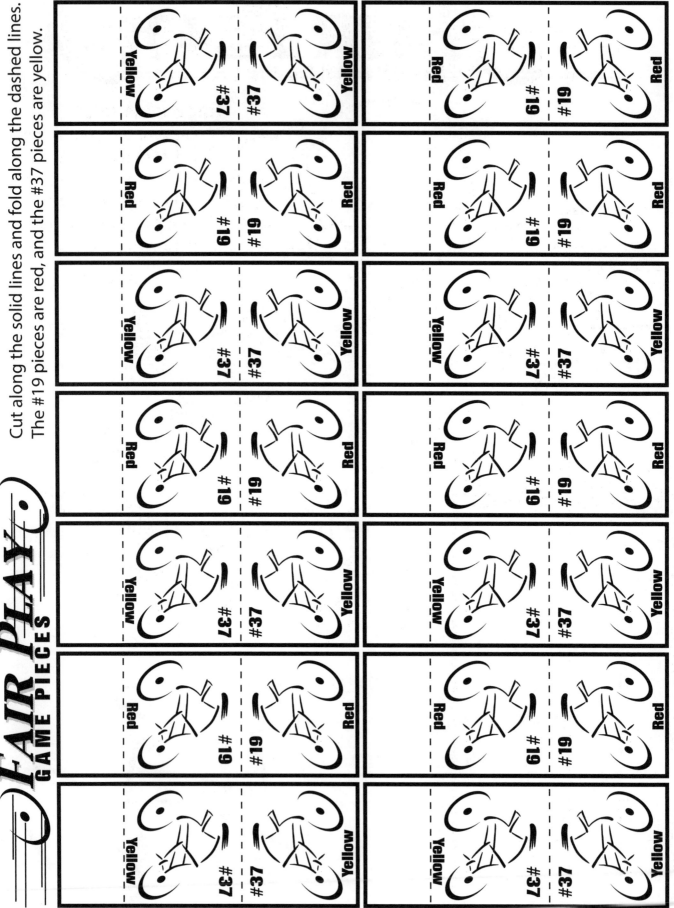

Cut along the solid lines and fold along the dashed lines.
The #19 pieces are red, and the #37 pieces are yellow.

FAIR PLAY
CHALLENGE RACE #1

Race #1 Rules:
1. On your turn, draw one cube from the bag.
2. Move forward one space if the cube color is the same as your bike color.
3. Return the cube to the bag.

START!

FINISH!

RACE	WINNER
1	
2	
3	
4	
5	
6	

1. Was the race fair? Do you think the riders had an equal chance of winning? Explain.

FAIR PLAY
CHALLENGE RACE #2

Race #2 Rules:

1. On your turn, draw two cubes from the bag.
2. Rider #19: Move forward one space if both cubes are the same color. Rider #37: Move forward one space if both cubes are different colors.
3. Return the cubes to the bag.

START!

FINISH!

RACE	WINNER
1	
2	
3	
4	
5	
6	

1. Was the race fair? Do you think the riders had an equal chance of winning? Explain.

FAIR PLAY
CLASS RESULTS

GROUP	CHALLENGE RACE #1 WINNER			CHALLENGE RACE #2 WINNER		
	#19	#37	Tie	#19	#37	Tie
1						
2						
3						
4						
5						
6						
7						
8						
9						
10						
11						
12						
13						
14						
15						
TOTALS						

Connecting Learning

1. In your group, who won the first race most often? Did you ever tie?

2. Based on our class results, did both players have an equal chance of winning?

3. If we had played the game 10 times instead of six, do you think the results would have been different? Why or why not?

4. Was the first race fair? Did both players have an equal chance of winning? Why or why not?

5. Describe the probable outcome of the first race using words like *likely*, *unlikely*, or *equally likely*.

6. In your group, who won the second race most? Did you ever tie? Did #19 ever win?

Connecting Learning

7. Based on our class results, did both players have an equal chance of winning the second race?

8. Was the second race fair? Why or why not?

9. Describe the probable outcome of the second race using words like *likely*, *unlikely*, or *equally likely*.

10. How does listing the possible outcomes help you see if a race is fair or not?

11. What would you do to make the second race fair?

SUM DOMINO DISCOVERIES

Topic
Probability

Key Question
What is the probability P(prime sum) when a domino is drawn randomly from a double-six set that the sum of the dots will be a prime number?

Learning Goals
Students will:
- determine the experimental probability of drawing a prime sum from a domino set, and
- determine the theoretical probability of drawing a prime sum from a domino set.

Guiding Documents
Project 2061 Benchmarks
- *How probability is estimated depends on what is known about the situation. Estimates can be based on data from similar conditions in the past or on the assumption that all the possibilities are known.*
- *Probabilities are ratios and can be expressed as fractions, percentages, or odds.*
- *The larger a well-chosen sample is, the more accurately it is likely to represent the whole, but there are many ways of choosing a sample that can make it unrepresentative of the whole.*
- *Events can be described in terms of being more or less likely, impossible, or certain.*

Common Core State Standards for Mathematics*
- *Make sense of problems and persevere in solving them. (MP.1)*
- *Reason abstractly and quantitatively. (MP.2)*
- *Construct viable arguments and critique the reasoning of others. (MP.3)*
- *Model with mathematics. (MP.4)*
- *Look for and make use of structure. (MP.7)*
- *Investigate chance processes and develop, use, and evaluate probability models. (7.SP.C)*

Math
Data analysis
  probability
  sampling
  data displays
    bar graph
Number sense
  prime numbers
  composite numbers

Integrated Processes
Observing
Classifying
Collecting and recording data
Organizing data
Comparing and contrasting
Predicting
Inferring

Materials
Double-six domino set
Bag or box to mix set
Student pages

Background Information
Experimental probability is determined by making a sufficiently large number of random draws from a double-six set of dominos. In each trial, the domino that is drawn is placed back into the box and the dominoes are thoroughly mixed to ensure randomness.

The theoretical probability is determined by (1) classifying the 28 dominoes by their sums, (2) determining how many have a sum that is prime, and (3) computing the probability using the formula: P(Prime Sum) = Number with prime sum/28.

As the number of draws increases, the experimental probability should come closer and closer to the theoretical probability.

When the dominoes are formed into a bar graph by stacking, an interesting pattern results. If sums are used, the dominoes form a regular symmetric rising and falling staircase pattern. When differences are used, the dominoes form a staircase that steps down as the difference increases, since there are a total of seven doubles that have a zero difference and only one that has a difference of six.

These patterns provide a rewarding surprise to most students since they are not acquainted with the patterns in dominoes. They also make it convenient for computing the various probabilities.

Management
1. Students should be familiar with prime and composite numbers. They also need to be aware that zero and one are neither prime nor composite, but are considered special cases.
2. This investigation works well for small groups of three or four students. Each can have a role: mixer, drawer, recorder, or stacker.

3. If sets of dominoes are not available for each group, copy the set provided onto card stock and cut them apart.
4. Caution students that they are to make the trials as random as possible. After a domino is drawn and the result is recorded, it should be placed back in the bag or box and the dominoes mixed thoroughly.
5. First, have each group complete its investigation and report the results. Then, have the students compile the composite of all the experimental results, thereby greatly increasing the total number of trials.
6. Some students find this investigation more motivating if it is presented in a game context. The *Key Question* might be restated: "Is it a fair game if I win when a composite is drawn and you win if it is a prime or special case?" The investigation can be extended by asking how the rules would need to be changed to make it fair.

Procedure
1. Distribute the dominoes, bags or boxes, and student pages.
2. Discuss the importance of getting random results and conclude that thorough shaking between draws and replacement will provide random results.
3. Have each group proceed through the investigation and determine the experimental probability.
4. Have students construct a graph with the dominoes and then make a representational bar graph to reflect the domino graph.
5. Referring to the graph, have students determine the theoretical probability.
6. Ask students to compare the experimental and theoretical probabilities and explain any differences.
7. Combine the experimental results of all groups to increase the number of trials. Compare the results of the larger sample with those of the smaller samples.
8. Have students explore other probabilities that can be determined from the information in the sum (and difference) column(s).
9. Have students share the experimental probabilities of the properties they selected.

Connecting Learning
1. What is the range of experimental probability results for prime sums among the class?
2. How did the probability results from the groups compare to the class composite?
3. What are the possible sums on a set of dominoes? [0-12] ...differences? [0-6]
4. What pattern do you see in the graph of dominoes? [sum: up and down staircase highest at 6, difference: down staircase highest at zero]
5. What sum is made by the greatest number of dominoes? [6] ...difference? [0]
6. What sum did your group get most often when doing the experiment? ...difference?
7. What reasons might there be for any differences between the experimental and the theoretical probabilities?
8. Would it be a fair game if I win when a composite number is drawn and you win if the number is a prime or special case? [No. More than half of the dominoes have composite sums, so composites win 15/28 of the time.]

Extensions
1. Do the activity *Different Domino Discoveries*. This investigation follows the same procedures as *Sum Domino Discoveries*. The student pages provide all the additional directions needed.
2. Have students develop a game of drawing dominoes that appears to be fair but turns out to be unfair (even vs. odd, six or less vs. seven or more). Have them play the game to determine how the probabilities work out experimentally and theoretically.

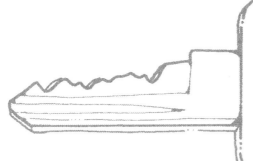

SUM DOMINO DISCOVERIES

SUM · DOMINO · DISCOVERIES

Key Question

What is the probability P(prime sum) when a domino is drawn randomly from a double-six set that the sum of the dots will be a prime number?

Learning Goals

Students will:

- determine the experimental probability of drawing a prime sum from a domino set, and
- determine the theoretical probability of drawing a prime sum from a domino set.

SUM DOMINO DISCOVERIES

Using a double-six set of dominoes, what is the probability of drawing a domino on which the sum of the dots is a prime number?

☐ What possible prime number sums can occur?

☐ What is your estimate of the probability of getting a prime sum (P(Prime Sum))?

To determine the experimental probability, draw 50 samples using the following procedure:

☐ Place the 28 dominoes in a bag or box.

☐ Mix thoroughly at the start and after each replacement.

☐ Draw out a domino and record the sum of the dots.

☐ Classify the result as prime, composite, or special case.

☐ Place the domino back in the bag or box.

We found the experimental probability for P(Prime Sum)=

Trial	Sum	Prime	Composite	Special Case
1.				
2.				
3.				
4.				
5.				
6.				
7.				
8.				
9.				
10.				
11.				
12.				
13.				
14.				
15.				
16.				
17.				
18.				
19.				
20.				
21.				
22.				
23.				
24.				
25.				
26.				
27.				
28.				
29.				
30.				
31.				
32.				
33.				
34.				
35.				
36.				
37.				
38.				
39.				
40.				
41.				
42.				
43.				
44.				
45.				
46.				
47.				
48.				
49.				
50.				
TOTAL				

SUM DOMINO DISCOVERIES

⚀ Build a graph by placing the dominoes with the same sum in one stack.
Arrange the stacks in order by sums of 0 to 12.

⚁ Complete the bar graph by coloring or shading in the number of dominoes in each of the stacks.

```
  0   1   2   3   4   5   6   7   8   9   10  11  12
```

⚂ From your graph, what is the theoretical probability for P(Prime Sum)= _____

⚃ Using the information in the graph and table, determine the experimental and theoretical probabilities for an even sum, an odd sum, and a sum of 7.

⚁ Description ⚁	Experimental Probability	Theoretical Probability
P(Even Sum)		
P(Odd Sum)		
P(Sum 7)		

⚄ Use the chart at the right to enter five other probabilities you could compare using the information in the table and graph.

⚁ Description ⚁	Experimental Probability	Theoretical Probability
P( )		
P( )		
P( )		
P( )		
P( )		

DIFFERENT DOMINO DISCOVERIES

Using a double-six set of dominoes, what is the probability of drawing a domino on which the difference between the two sets of dots is a prime number?

- What possible prime number differences can occur?

- What is your estimate of the probability of getting a prime difference (P (Prime Difference))

To determine the experimental probability, draw 50 samples using the following procedure:

- Place the 28 dominoes in a bag or box.

- Mix thoroughly at the start and after each replacement.

- Draw out a domino and record the difference of dots.

- Classify the result as prime, composite, or special case.

- Place the domino back in the bag or box.

We found the experimental probability for P(Prime Difference)=

Trial	Difference	Prime	Composite	Special Case
1.				
2.				
3.				
4.				
5.				
6.				
7.				
8.				
9.				
10.				
11.				
12.				
13.				
14.				
15.				
16.				
17.				
18.				
19.				
20.				
21.				
22.				
23.				
24.				
25.				
26.				
27.				
28.				
29.				
30.				
31.				
32.				
33.				
34.				
35.				
36.				
37.				
38.				
39.				
40.				
41.				
42.				
43.				
44.				
45.				
46.				
47.				
48.				
49.				
50.				
TOTAL				

DIFFERENT · DOMINO · DISCOVERIES

DIFFERENT DOMINO DISCOVERIES

⚀ Build a graph by placing the dominoes with the same difference in one stack. Arrange the stacks in order by differences of 0 to 6.

⚁ Complete the bar graph by coloring or shading in the number of dominoes in each of the stacks.

```
0   1   2   3   4   5   6
```

⚂ From your graph, what is the theoretical probability for P(Prime Difference)= _____

⚃ Using the information in the graph and table, determine the experimental and theoretical probabilities for an even difference, an odd difference, and a difference of 2.

⚂ Description ⚂	Experimental Probability	Theoretical Probability
P(Even Difference)		
P(Odd Difference)		
P(Difference of 2)		

⚄ Use the chart at the right to enter five other probabilities you could compare using the information in the table and graph.

⚀ Description ⚀	Experimental Probability	Theoretical Probability
P( )		
P( )		
P( )		
P( )		
P( )		

SUM DOMINO DISCOVERIES

SUM DOMINO DISCOVERIES

SUM DOMINO DISCOVERIES

Connecting Learning

1. What is the range of experimental probability results for prime sums among the class?

2. How did the probability results from the groups compare to the class composite?

3. What are the possible sums on a set of dominoes? ...difference?

4. What pattern do you see in the graph of dominoes?

Connecting Learning

5. What sum is made by the greatest number of dominoes? ...difference?

6. What sum did your group get most often when doing the experiment? ...differences?

7. What reasons might there be for any differences between the experimental and theoretical probabilities?

8. Would it be a fair game if I win when a composite number is drawn and you win if the number is a prime of special case?

Dicey Decisions

Topic
Probability

Key Question
Which die should you pick to roll the highest number most of the time?

Learning Goals
Students will:
- play a game using four non-standard dice,
- evaluate the experimental and theoretical probabilities associated with the game, and
- use what they learned to develop a strategy that allows them to win most of the time.

Guiding Documents
Project 2061 Benchmarks
- *Estimate probabilities of outcomes in familiar situations, on the basis of history or the number of possible outcomes.*
- *Probabilities are ratios and can be expressed as fractions, percentages, or odds.*
- *Events can be described in terms of being more or less likely, impossible, or certain.*

Common Core State Standards for Mathematics*
- *Make sense of problems and persevere in solving them. (MP.1)*
- *Reason abstractly and quantitatively. (MP.2)*
- *Construct viable arguments and critique the reasoning of others. (MP.3)*
- *Use appropriate tools strategically. (MP.5)*
- *Investigate chance processes and develop, use, and evaluate probability models. (7.SP.C)*

Math
Data analysis
    probability
Logical thinking

Integrated Processes
Observing
Comparing and contrasting
Collecting and recording data
Interpreting
Analyzing
Generalizing
Applying

Materials
Wooden cubes (see *Management 1*)
Student pages

Background Information
This activity originated with a Stanford University statistician, Bradley Efron, who designed the game as a study in probability. Using four non-standard dice that they construct themselves, students will play a game that involves both careful logical consideration and a little bit of luck.

The object of the game is to roll the highest number. One student begins by choosing one of the four non-standard dice. The second student then chooses a die and both roll. The highest roll wins. Six rolls are made using the same set of dice; each roll is recorded in one of the tables on the third student page. For the next game, the students switch who chooses first and repeat the process.

What becomes apparent after playing several games is that the person choosing first is at a distinct disadvantage. No matter which die they choose, the person choosing second can select a die with a two-thirds probability of winning. Once students realize this, only a series of less probable rolls will make the person who chooses second lose.

To see the reasons for this, every possible combination of two dice must be considered. When the probabilities for a given die winning in each combination are examined, it becomes clear that one die is always at a distinct advantage.

Management
1. If possible, use wooden cubes or other uniform cube shapes to make the non-standard dice. Students can write the numbers on the faces as instructed on the first student page. Another option is to cover the faces of standard dice with self-adhesive dots on which the revised numbers are written. Each pair of students will need four cubes.
2. This activity is designed to be played in pairs. Each set of two students will need one copy of the first two student pages and one set of non-standard dice.
3. This activity is divided into two parts. *Part One* introduces the game and has students look at experimental probability. *Part Two* has students analyze the theoretical probabilities of the game. These parts can be done on the same day or on two different days.

Procedure

Part One

1. Have students get into pairs and distribute the first student page and the materials for constructing the non-standard dice to each group.
2. Once groups have constructed their sets of four dice, hand out the remaining *Part One* student pages—one copy for each student.
3. Go over the rules and be sure students understand the procedure. Allow time for them to play six games and record their rolls and who won in each case.
4. After playing six games, students should begin to develop some strategies for winning the most often. The table and questions on the last two student pages should help students recognize and verbalize these strategies.
5. Before answering the questions on the final student page, have students play at least four more games to test their strategies. (In these subsequent games, students do not need to roll six times with the same set of dice, but they should be sure to switch who chooses first each game.)
6. Discuss students' strategies and what they learned about probability.

Part Two

1. Distribute the first two student pages for *Part Two*. Go over the instructions and be sure that everyone understands how to complete the grids.
2. Have students get back into their groups and give them time to complete all five grids.
3. When groups are done, distribute the remaining two student pages. Allow time for students to discuss the theoretical probabilities and answer the questions.
4. Close with a final time of class discussion where students compare their experimental probabilities with the theoretical ones.

Connecting Learning

Part One

1. Who won more often in the first six games you played—the person choosing first or the person choosing second?
2. Did this change once you developed a strategy? Why or why not?
3. Do you think it matters if you choose first or second? [Yes.] Why or why not? [If you choose second, you can always select a die that will give you a greater chance of winning when paired with the die chosen by the first person.]
4. Describe your chances of winning the game if you choose your die first. [Unlikely if the person choosing second selects properly.] ...if you choose second. [Likely if the correct die is selected.]

5. Is it impossible for the person who chooses first to win? Why or why not?
6. If the person going first chose Die A, which die would you choose? Why?
7. What if he or she chose Die B? ...Die C? ...Die D? Justify your responses.
8. Describe the strategy you developed for choosing your die each game.

Part Two

1. Would the way you play the game change now that you have analyzed the theoretical probabilities? Why or why not?
2. What strategy would you use if you always had to pick the first die? Why?
3. In the games that you played, did the die with the theoretical probability of winning always win? Why or why not?
4. Is it possible for Die B to beat Die A? [Yes.] Is it likely? [Not very likely. Die B will win 2/3 of the time according to the theoretical probabilities.]
5. If you could pick any combination of two dice to give yourself the best chance of winning, which two would you pick? Which of those two dice would you want to have? Why?

Dicey Decisions

Key Question
Which die should you pick to roll the highest number most of the time?

Learning Goals

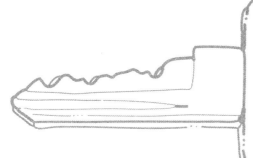

Students will:

- play a game using four non-standard dice,
- evaluate the experimental and theoretical probabilities associated with the game, and
- use what they learned to develop a strategy that allows them to win most of the time.

Dicey Decisions Part One

Getting Started:

With your partner, make a set of four dice. Label the sides of the dice with the following numbers:

Die A:
0: two sides
4: four sides

Die B:
3: all six sides

Die C:
2: four sides
6: two sides

Die D:
1: three sides
5: three sides

Playing the Game:

This is a game of chance where you must try to make the odds go in your favor.

1. Assign one person to be Player A and one person to be Player B.
2. Player A begins by choosing one of the four dice. Player B chooses one of the remaining three dice.
3. Record the die chosen by each player on the recording page.
4. Each player rolls his or her die, and the person with the highest number wins. Record the letter of the winning die in the *Winner* column.
5. Roll both dice six times, recording who won each roll. This is one complete game.
6. At the bottom of the table, record the number of times each player won in that game.
7. Play six games. Each game, trade which player chooses his or her die first.
8. At the bottom of the page, record the number of games each player won overall.

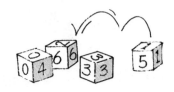

Dicey Decisions
Part One

Follow the instructions on the previous page. Play the game six times. Record the results of each game. Remember to rotate which player picks first.

Player A's die: _____
Player B's die: _____

Game One

Roll	Winner
1	
2	
3	
4	
5	
6	

Player A won _____ times
Player B won _____ times

Player B's die: _____
Player A's die: _____

Game Two

Roll	Winner
1	
2	
3	
4	
5	
6	

Player A won _____ times
Player B won _____ times

Player A's die: _____
Player Bs die: _____

Game Three

Roll	Winner
1	
2	
3	
4	
5	
6	

Player A won _____ times
Player B won _____ times

Player B's die: _____
Player A's die: _____

Game Four

Roll	Winner
1	
2	
3	
4	
5	
6	

Player A won _____ times
Player B won _____ times

Player A's die: _____
Player B's die: _____

Game Five

Roll	Winner
1	
2	
3	
4	
5	
6	

Player A won _____ times
Player B won _____ times

Player B's die: _____
Player A's die: _____

Game Six

Roll	Winner
1	
2	
3	
4	
5	
6	

Player A won _____ times
Player B won _____ times

Player A won _____ games overall. Player B won _____ games overall.

There were _____ games that were ties.

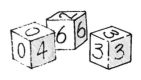

Using the information from the tables on the previous page, fill in the table below to compare the number of times each die won in the different games you played. Starting with *Game One*, record the number of times each die won out of the total number of rolls in the game. This number should be represented as a fraction such as 2/6 or 5/6.

Game	Die	Wins/Total
One		/6
One		/6
Two		/6
Two		/6
Three		/6
Three		/6
Four		/6
Four		/6
Five		/6
Five		/6
Six		/6
Six		/6

Look at the results of your first six games in the table. Use what you see and what you noticed while you were playing to develop a strategy that will help you win most of the time. Try to use this strategy as you play the game a few more times with your partner.

Dicey Decisions Part One

1. In the first six games you played, who won more often, the person choosing first or the person choosing second? Why?

2. In the games you played after you developed a strategy, who won more often, the person choosing first or the person choosing second? Why?

3. Do you think it matters if you choose first or second? Why or why not?

4. If the person going first chose Die A, which die would you choose? Why?

5. What if he or she chose Die B? …Die C? …Die D? Justify your responses.

6. Please describe the strategy you developed for choosing your die.

Dicey Decisions Part Two

These grids list the numbers on the faces of two dice. They are one way to look at all of the possible combinations that can occur when rolling two dice together. Each square in the grid represents a unique roll that is possible when the two dice listed are paired.

Determine the winner for each possible roll. Write the letter of the winning die in the square. Record how many times each die wins out of the 36 possible rolls. The first grid has been completed for you as an example.

Die A

	0	0	4	4	4	4
3	B	B	A	A	A	A
3	B	B	A	A	A	A
3	B	B	A	A	A	A
3	B	B	A	A	A	A
3	B	B	A	A	A	A
3	B	B	A	A	A	A

Die B (left side label)

Die A wins 24 times out of 36.

Die B wins 12 times out of 36.

Die A

	0	0	4	4	4	4
2						
2						
2						
2						
6						
6						

Die C (left side label)

Die A wins ____ times out of 36.

Die C wins ____ times out of 36.

Complete these tables like you did the one on the previous page.

Die A

	0	0	4	4	4	4
1						
1						
1						
5						
5						
5						

(left axis: **Die D**)

Die A wins ____ times out of 36.

Die D wins ____ times out of 36.

Die B

	3	3	3	3	3	3
2						
2						
2						
2						
6						
6						

(left axis: **Die C**)

Die B wins ____ times out of 36.

Die C wins ____ times out of 36.

Die B

	3	3	3	3	3	3
1						
1						
1						
5						
5						
5						

(left axis: **Die D**)

Die B wins ____ times out of 36.

Die D wins ____ times out of 36.

Die C

	2	2	2	2	6	6
1						
1						
1						
5						
5						
5						

(left axis: **Die D**)

Die C wins ____ times out of 36.

Die D wins ____ times out of 36.

Dicey Decisions Part Two

Use the values from the tables on the previous pages to fill in the table of theoretical probabilities below. Probabilities are often written as fractions. For example, to say that the probability of rolling a one on Die D is three out of six we would write P(1) = 3/6. Simplified to lowest terms, that would be P(1) = 1/2.

For each combination, write the total number of times each die would win over the total number of possible combinations. Simplify each fraction to its lowest terms. The first combination has been done for you as an example.

Combination	Probability of winning	
A, B	P(A) = 24/36 = 2/3	P(B) = 12/36 = 1/3
A, C	P(A) =	P(C) =
A, D	P(A) =	P(D) =
B, C	P(B) =	P(C) =
B, D	P(B) =	P(D) =
C, D	P(C) =	P(D) =

 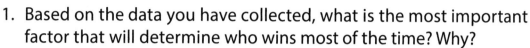

Dicey Decisions Part Two

1. Based on the data you have collected, what is the most important factor that will determine who wins most of the time? Why?

2. How does knowing the probabilities that a given die will win change the way you play this game?

3. Outline the strategy you would use if you always had to pick first. Defend your reasoning.

4. How do the theoretical probabilities compare to the actual results you got when playing this game?

5. What are some factors that could account for these differences?

Dicey Decisions

Connecting Learning

Part One

1. Who won more often in the first six games you played—the person choosing first or the person choosing second?

2. Did this change once you developed a strategy? Why or why not?

3. Do you think it matters if you choose first or second? Why or why not?

4. Describe your chances of winning the game if you choose your die first ...if you choose second.

5. Is it impossible for the person who chooses first to win? Why or why not?

6. If the person going first chose Die A, which die would you choose? Why?

7. What if he or she chose Die B? ...Die C? ...Die D? Justify your responses.

Dicey Decisions

Connecting Learning

8. Describe the strategy you developed for choosing your die each game.

Part Two

1. Would the way you play the game change now that you have analyzed the theoretical probabilities? Why or why not?

2. What strategy would you use if you always had to pick the first die? Why?

3. In the games that you played, did the die with the theoretical probability of winning always win? Why or why not?

4. Is it possible for Die B to beat Die A? Is it likely?

5. If you could pick any combination of two dice to give yourself the best chance of winning, which two would you pick? Which of those two dice would you want to have? Why?

Playing for ODDS

Topic
Probability

Key Question
In this dice game, I win when the product of the roll is even and you win when the product of the roll is odd. How can you determine if this is a "fair" game?

Learning Goals
Students will:
- determine the experimental probability of throwing two dice that have a product that is an odd number,
- determine the theoretical probability of throwing two dice that have a product that is an odd number, and
- evaluate whether or not a game in which the player rolling an odd product wins is fair to both players.

Guiding Documents
Project 2061 Benchmarks
- *How probability is estimated depends on what is known about the situation. Estimates can be based on data from similar conditions in the past or on the assumption that all the possibilities are known.*
- *Probabilities are ratios and can be expressed as fractions, percentages, or odds.*
- *The larger a well-chosen sample is, the more accurately it is likely to represent the whole, but there are many ways of choosing a sample that can make it unrepresentative of the whole.*
- *Events can be described in terms of being more or less likely, impossible, or certain.*

Common Core State Standards for Mathematics*
- *Make sense of problems and persevere in solving them. (MP.1)*
- *Reason abstractly and quantitatively. (MP.2)*
- *Construct viable arguments and critique the reasoning of others. (MP.3)*
- *Use appropriate tools strategically. (MP.5)*
- *Investigate chance processes and develop, use, and evaluate probability models. (7.SP.C)*

Math
Data analysis
  probability
  sampling

Integrated Processes
Observing
Classifying
Collecting and recording data
Organizing data
Comparing and contrasting
Predicting
Inferring

Materials
Dice, 2 per group
Student pages

Background Information
In this activity, experimental probability is determined by making a sample of random rolls of a pair of dice and finding the fraction of the sample that has odd products. The theoretical probability is determined by (1) classifying the 36 possible outcomes by their products, (2) determining how many have a product that is odd, and (3) computing the probability using the formula: P(Odd Product) = Number with odd products/36. As the number of rolls increases, the experimental probability should come closer and closer to the theoretical probability.

Making a grid of all the possible outcomes with the products is an appropriate way to find all the possible outcomes. This grid forms the familiar multiplication table. As students shade in the odd products, the pattern that every other row and column is even becomes evident. A quick count shows that only nine out of 36 products are the winning odd product. The equivalent ratios express why an odd product is not a favorable outcome, 9/36, 1/4, 25%.

The grid provides a useful tool in determining other probabilities to develop a game that is more equitable to both players. Such might include products greater than or equal to 18, P(product ≥ 18) = 10/36 ≈ .28 = 28%, or products greater than or equal to 12, P(product ≥ 12) = 17/36 ≈ .47 = 47%.

Management

1. This investigation works well for pairs of students so each can roll a die.
2. First, have each group complete their investigation and report the results. Then have the students compile the composite of all the experimental results, thereby greatly increasing the total number of trials.

Procedure

1. Introduce the lesson by explaining the rules of this game to the class and asking the *Key Question*. Encourage the class to recognize the apparent fairness of the game by noting that of the six numbers on a die, three are odd and three are even.
2. Distribute the dice and the first student page.
3. Have each pair of students make 20 rolls and determine the experimental probability of rolling two numbers whose product is odd. Encourage them to evaluate the fairness of the game based on their initial data.
4. Distribute the second student page. As a class, combine the total number of odd products and total number of rolls to determine the experimental probability of a larger sample. Have students evaluate the fairness of the game based on this larger sample of data.
5. Give students the third page and have them complete the grid to determine all the possible products.
6. Referring to the grid, have students determine the theoretical probability of rolling two numbers whose product is odd.
7. Ask students to compare the experimental and theoretical probabilities and explain any differences.
8. Referring to the grid, have students explore other probabilities to determine a set of rules that would be more fair to each player.
9. Have students share the experimental probabilities for the modified rules they selected.

Connecting Learning

1. When you and your partner collected data, did the game appear to be fair? Why or why not?
2. How did the probability results from the groups compare to the class composite?
3. What is the range of experimental probability results for odd products among the class?
4. Based on the data from the entire class, do you think the game is fair? Why or why not?
5. What pattern do you see in the grid of possible products? [even products every other row and column; only nine odd products; rows and columns are counting by ones, twos, threes, etc.; looks like the multiplication table; square numbers on descending left to right diagonal; symmetry across square number diagonal; etc.]
6. What products show up more than any others in the grid? [6 and 12]
7. What products did your group get most often when doing the experiment?
8. Did your group or the class data come closer to matching the theoretical probability? Why?
9. What reasons might there be for any differences between the experimental and theoretical probabilities?
10. What rules did you devlelop for a game that would be more fair to both players? What is the probability of winning using your new rules?

Extension

Have students develop a game based on rolling dice that appears to be fair but turns out to be unfair (sums and differences work well). Have them play the game to determine how the probabilities work out experimentally and theoretically.

Playing for ODDS

Learning Goals

Students will:

- determine the experimental probability of throwing two dice that have a product that is an odd number,
- determine the theoretical probability of throwing two dice that have a product that is an odd number, and
- evaluate whether or not a game in which the player rolling an odd product wins is fair to both players.

Playing for ODDS

When a pair of dice is rolled and the two numbers are multiplied, the product ranges from 1 to 36. In this game, one player wins if the product is odd, and the other player wins if the product is even.

With a partner, roll a pair of dice 20 times, record the outcomes and products, and determine the winner. Use this data to decide if the game is fair.

How many times was an odd product rolled?

What fraction of the rolls were odd products?

Write the fraction of odd products as a decimal.

What percent of the rolls were odd products?

Based on your data, is this game fair? Why or why not?

Trial	Roll of Dice		Winner	
	Die 1 · Die 2 = Product		Odd	Even
1.	· =			
2.	· =			
3.	· =			
4.	· =			
5.	· =			
6.	· =			
7.	· =			
8.	· =			
9.	· =			
10.	· =			
11.	· =			
12.	· =			
13.	· =			
14.	· =			
15.	· =			
16.	· =			
17.	· =			
18.	· =			
19.	· =			
20.	· =			
	Final Score			

Group Data

P(odd product)

Odd Rolls	
Total Rolls	
Fraction	
Decimal	
Percent	

Playing for ODDS

Maybe you and your partner just had a bad set of rolls. Gather data from the whole class to see if it matches your data.

How many odd products were rolled?

What was the total number of rolls made in the class?

What fraction of the rolls were odd products?

Write the fraction of odd products as a decimal.

What percent of the rolls were odd products?

Based on the class data, is this game fair? Why or why not?

Group Number	Number of Odd Products	Number of Rolls
1.		
2.		
3.		
4.		
5.		
6.		
7.		
8.		
9.		
10.		
11.		
12.		
13.		
14.		
15.		
16.		
17.		
18.		
Class Total		

P(odd product)

Class Data

Odd Rolls	
Total Rolls	
Fraction	
Decimal	
Percent	

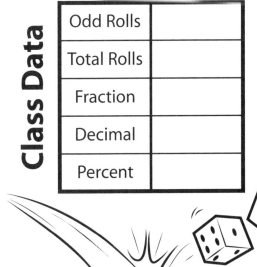

Playing for ODDS

Determine all the possible products by filling in the grid with the product of every combination. Then shade in the odd products.

Determine the theoretical probability of getting an odd product.

How many odd products are there?

What fraction of the products is odd?

What percent of the products are odd?

Die One

	1	2	3	4	5	6
1						
2						
3						
4						
5						
6						

Die Two

How well did the class data match this percent?

How well did you and your partner's data match the percent?

Description	Experimental Probability	Theoretical Probability
P(odd product)	Class Data	
P(odd product)	Group Data	

Change the rules to make a game that is more fair. Use the data from your grid to determine the probability of winning.

LET'S CHANGE THIS UP TO MAKE IT MORE FAIR, PARDNER!

Connecting Learning

1. When you and your partner collected data, did the game appear to be fair? Why or why not?

2. How did the probability results from the groups compare to the class composite?

3. What is the range of experimental probability results for odd products among the class?

4. Based on the data from the entire class, do you think the game is fair? Why or why not?

5. What pattern do you see in the grid of possible products?

Connecting Learning

6. What products show up more than any others in the grid?

7. What products did your group get most often when doing the experiment?

8. Did your group or the class data come closer to matching the theoretical probability? Why?

9. What reasons might there be for any differences between the experimental and theoretical probabilities?

10. What rules did you develop for a game that would be more fair to both players? What is the probability of winning using your new rules?

Probability on a Roll

Topic
Probability

Key Questions
1. What is the average number of rolls it takes to finish the game?
2. What are the probabilities of finishing in different numbers of rolls?

Learning Goals
Students will:
- play a game multiple times;
- keep track of the number of rolls it takes them to get from start to finish;
- find the individual, group, and class average number of rolls it takes to complete the game; and
- look at the experimental probabilities of finishing in each possible number of rolls.

Guiding Documents
Project 2061 Benchmarks
- *Add, subtract, multiply, and divide whole numbers mentally, on paper, and with a calculator.*
- *Find the mean and median of a set of data.*
- *Estimate probabilities of outcomes in familiar situations, on the basis of history or the number of possible outcomes.*
- *Events can be described in terms of being more or less likely, impossible, or certain.*

Common Core State Standards for Mathematics*
- *Make sense of problems and persevere in solving them. (MP.1)*
- *Reason abstractly and quantitatively. (MP.2)*
- *Construct viable arguments and critique the reasoning of others. (MP.3)*
- *Investigate chance processes and develop, use, and evaluate probability models. (7.SP.C)*

Math
Data analysis
    probability
    measures of central tendency
      mean

Integrated Processes
Observing
Collecting and recording data
Comparing and contrasting
Analyzing
Generalizing

Materials
For each group:
    game board
    one die

For each student:
    game piece (see *Management 2*)
    student pages

For the class:
    transparency of frequency chart

Background Information
This activity presents a game whose object is for players to take turns rolling the die and moving their game pieces the corresponding number of spaces until they reach or exceed the *Finish* space. The emphasis is on keeping track of how many rolls it takes each player to do this and then to compare and contrast these values to each other, as well as to the maximum and minimum number of rolls possible. Because there are 24 spaces on the game board (not including the *Start*), it is theoretically possible to move from the start to the finish in four rolls (by rolling a six each time); however, this is unlikely. Likewise, it is possible, though extremely unlikely, that it could take a player 24 rolls to complete the journey (by rolling a one each time).

As students play multiple times, they will collect and organize data and use that data to determine averages for individuals, groups, and the entire class. They will then assign a description of probability (impossible, very unlikely, unlikely, likely, very likely, certain) to each of eight scenarios. In order to do this, students must know the experimental probability of finishing the game in each possible number of rolls. The experimental probability reflects the actual results of students' games, and may or may not be similar to the theoretical probabilities. For example, the theoretical probability of completing the game in four rolls is 1/1296. [The chance of rolling one six is 1/6, the chance of rolling two sixes in a row is 1/36

(1/6 x 1/6), the chance of rolling three in a row is 1/216 (1/6 x 1/6 x 1/6), and the chance of rolling four in a row is 1/1296 (1/6 x 1/6 x 1/6 x 1/6).] It is very unlikely that any student will finish the game in four rolls, but it is possible. It is also unlikely that any student will finish the game in five rolls (theoretical probability: 1/864), but this outcome is more likely than the four-rolls outcome.

Management

1. The activity is designed as a game to be played in groups of two to four.
2. Any small manipulatives can be used as game pieces as long as there is a way to distinguish between the pieces used by the members of one group (different colors of Teddy Bear Counters, different coins, different buttons, etc.).

Procedure

1. Have students get into their groups and distribute the game board, die, game pieces, and first two student pages.
2. Before beginning to play, have students analyze the game within their groups so that they are able to recognize and verbalize the "limits" to the number of rolls possible (see *Background Information*). Be sure that each student records his or her guess as to how many rolls it will take to finish the first game.
3. Instruct students to take turns rolling the die and moving their marking chips the appropriate number of spaces. For each roll of the die, they should make one tally mark in the space provided on the first student page. No matter who finishes first, all players continue until they reach or exceed the *Finish* space.
4. Once a single round of play has been completed, take some time as a class to discuss how the actual results compared to the guesses that students made. Discuss what may have made some students over- or under-estimate the number of rolls it would take them to finish.
5. Have students make new predictions as to how many rolls it will take them to finish next time, then play another round and compare the predictions to the results once again.
6. After three or more rounds have been played, ask students to determine their individual and group averages using the tables on the second student page. Once all groups have determined the average number of rolls it took each player to finish, compile this information to determine the whole-class average.
7. Discuss whether or not this average is close to what should be expected. To do this, divide the number of spaces on the board (24) by the average number of rolls it took students to complete the game (we will use 7.5 as an example). This calculation gives you the average value rolled by all students in all games (24 ÷ 7.5 = 3.2). This average value rolled should be fairly close to 3.5 because that is the average of the numbers on a die (1 + 2 + 3 + 4 + 5 + 6 = 21; 21 ÷ 6 = 3.5) and there is an equal chance of rolling each number. If the average value rolled is significantly greater or less than 3.5, you may want to check for recording errors.

8. Put the transparency copy of the frequency chart up on the overhead. In the *Frequency* column, record the number of times the game was completed in each number of rolls. In the *Experimental Probability* column, put the frequency over the total number of games played. For example, if two students each finished one game in five rolls, and you had 30 students who each played the game four times, the experimental probability for five rolls would be 2/120.
9. Distribute the final student page and have students use the information from the chart to answer the questions.

Connecting Learning

1. What is the fewest number of rolls it could take to finish the game? [four] What would you have to roll in order for this to happen? [four sixes]
2. What is the greatest number of rolls it could take to finish the game? [24] What would you have to roll in order for this to happen? [24 ones]
3. How many rolls did you think it would take you to finish the game? Why did you guess this?
4. How close was your guess to your actual results? Did this change the second time you played the game? Why or why not?
5. What was the average number of rolls it took for your group to finish the game? How did this average compare to the averages of other groups?
6. Did anyone finish the game in four rolls? ...five rolls? ...more than 10 rolls? (If yes) Was this very likely? Why or why not?
7. What are you wondering now?

Key Questions

1. What is the average number of rolls it takes to finish the game?
2. What are the probabilities of finishing in different numbers of rolls?

Learning Goals

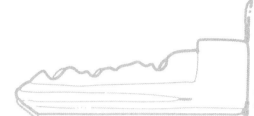

- play a game multiple times;

- keep track of the number of rolls it takes them to get from start to finish;

- find the individual, group, and class average number of rolls it takes to complete the game; and

- look at the experimental probabilities of finishing in each possible number of rolls.

203

Probability on a Roll

START

You're on a Roll!

Quick Sip
WATER STATION

FINISH

204

Probability on a Roll

PART 1

1. Look at the game board. Using a single die and moving one space for each number rolled, what is the fewest number of rolls it could take to finish the game? What would you have to roll in order for this to happen?

2. What is the greatest number of rolls it could take to finish the game? What would you have to roll in order for this to happen?

3. How many rolls do you think it will actually take you to finish the game? Why?

PART 2

Play one round of the game with your group. Keep a tally of the number of rolls it takes you to finish in the box below.

• **Number of Rolls Tally** •

4. Does the actual number of rolls match your prediction? Why or why not?

5. Make a revised prediction for the next round and write it here.

Probability on a Roll

After you have played at least three rounds with your group, determine the average number of rolls it took each person to finish the game.

PART 3

Number of Rolls

Player	Round 1	Round 2	Round 3	Round 4	Average

Combine these averages to come up with a group average.

Average Player 1	Average Player 2	Average Player 3	Average Player 4	Average Group

1. How does this average compare to your first guess about how many rolls it would take you to finish the game? …your second guess?

2. Combine the group averages to come up with a class average. How does this average compare to your individual and group averages?

3. If you and your classmates were to play the game 1000 times, do you think the average would go up, go down, or stay the same? Justify your response.

Probability on a Roll

Look at the experimental probabilities that you and your classmates generated by playing the game. Use one of the following words to answer each of the questions below:

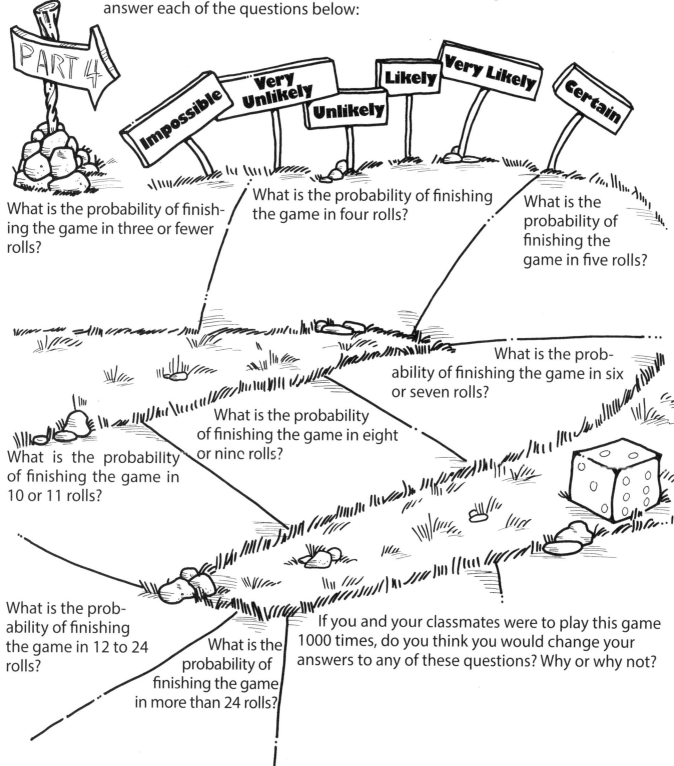

PART 4

Impossible Very Unlikely Unlikely Likely Very Likely Certain

What is the probability of finishing the game in three or fewer rolls?

What is the probability of finishing the game in four rolls?

What is the probability of finishing the game in five rolls?

What is the probability of finishing the game in six or seven rolls?

What is the probability of finishing the game in eight or nine rolls?

What is the probability of finishing the game in 10 or 11 rolls?

What is the probability of finishing the game in 12 to 24 rolls?

What is the probability of finishing the game in more than 24 rolls?

If you and your classmates were to play this game 1000 times, do you think you would change your answers to any of these questions? Why or why not?

Probability on a Roll

Number of rolls	Frequency	Experimental Probability
4		
5		
6		
7		
8		
9		
10		
11		
12		
13		
14		
15		
16		
17		
18		
19		
20		
21		
22		
23		
24		

Connecting Learning

1. What is the fewest number of rolls it could take to finish the game? What would you have to roll in order for this to happen?

2. What is the greatest number of rolls it could take to finish the game? What would you have to roll in order for this to happen?

3. How many rolls did you think it would take you to finish the game? Why did you guess this?

4. How close was your guess to your actual results? Did this change the second time you played the game? Why or why not?

5. What was the average number of rolls it took for your group to finish the game? How did this average compare to the averages of other groups?

6. Did anyone finish the game in four rolls? ...five rolls? ...more than 10 rolls? Was this very likely? Why or why not?

7. What are you wondering now?

Probability on a Roll, Again

Topic
Probability

Key Questions
1. What is the experimental probability of landing on each space on the game board?
2. How can you use the adaptation pieces to change the average number of rolls it takes to complete the game?

Learning Goals
Students will:
- determine the experimental probability of landing on each space of the game board,
- use adaptation pieces to modify the game board in specific ways,
- determine the average number of rolls it takes to complete the modified game, and
- use two of the adaptation pieces to make a game board that takes more (or fewer) rolls on average to complete than the original board.

Guiding Documents
Project 2061 Benchmarks
- *Add, subtract, multiply, and divide whole numbers mentally, on paper, and with a calculator.*
- *Find the mean and median of a set of data.*
- *Estimate probabilities of outcomes in familiar situations, on the basis of history or the number of possible outcomes.*

Common Core State Standards for Mathematics*
- *Make sense of problems and persevere in solving them. (MP.1)*
- *Reason abstractly and quantitatively. (MP.2)*
- *Construct viable arguments and critique the reasoning of others. (MP.3)*
- *Use appropriate tools strategically. (MP.5)*
- *Investigate chance processes and develop, use, and evaluate probability models. (7.SP.C)*

Math
Data analysis
    probability
    measures of central tendency
        mean

Integrated Processes
Observing
Collecting and recording data
Comparing and contrasting
Analyzing
Generalizing

Materials
For each group:
    2 game boards from *Probability on a Roll*
    one die
    adaptation pieces
    student pages

For each student:
    game piece (see *Management 1*)

Background Information
    This activity is an extension to the activity *Probability on a Roll* and takes a more in-depth look at concepts first explored there. It is divided into four parts. In *Part One*, students will obtain experimental data relating to the average number of rolls that it takes to complete the game as well as the probability of landing on any given space on the boar.

    The challenge in *Part Two* is for students to use logical thinking combined with what they learned from the experimental probabilities to solve problems. They are given five adaptation pieces. These pieces say things like "Double your next roll" or "Move back three spaces." Using one piece at a time, students are challenged to increase the minimum number of moves required to finish a game, decrease the minimum number of moves required, increase the maximum number of moves required, and decrease the maximum number of moves required. Each of these challenges can be solved in multiple ways, and some can be solved using several of the pieces. A single piece may accomplish two of the challenges at once, depending on where it is placed.

    Part Three asks students to select one of the modified game boards developed in *Part Two* and predict whether the average number of rolls it takes to complete the modified game will be higher, lower, or the same as the average from the original game board. Students will then play several times using the modified board to determine the average number of rolls it takes to complete the game. They are then to compare the new average with the average calculated for the original game board in *Part One*.

The fourth and final part of this activity asks students to combine what they learned about the experimental probabilities in *Part One* with what they determined about the adaptation pieces in *Part Two* to design their own game boards. They must use two of the adaptation pieces to either increase or decrease the average number of rolls it takes to complete the game. They will then determine the success of their design by playing several rounds of the game and calculating the averages for individuals and the group. These averages will be compared to the averages from *Part One* to determine if students were successful.

Management

1. This activity is a continuation of the activity *Probability on a Roll*. Students should complete that activity before moving on to this one. They will also need the game board, dice, and game pieces used in that activity. Students should remain with the same groups from *Probability on a Roll*.
2. This activity is divided into four parts and is designed to take place over several days. Depending on the needs and abilities of your students, you may wish to complete only some of the parts.
3. *Part One* can be done over several days so that students do not have to play 10 rounds of the game at one sitting. For this part of the activity, each group will need one copy of the first student page on which to keep track of the number of rolls and two copies of the game board from *Probability on a Roll*. One copy will be used to play the game, and the other will be used to record tallies each time a space is landed on. **Note:** If a player lands exactly on the *Finish*, he or she should record a tally in that space. If a player's final roll causes him or her to exceed the *Finish*, a tally should not be marked in that space.
4. Copy the page of adaptation pieces onto card stock. Each group needs one set of the pieces. (There are five pieces to a set.)

Procedure

Part One
1. Have students get into their same groups from *Probability on a Roll*. Distribute one copy of the first student page to each group. Give each group two copies of the game board.
2. Allow time (possibly over several days) for groups to play the game 10 times and record their data.
3. When groups have collected at least 10 games worth of data, have them determine the experimental probability of landing on each space. This probability is the number of times each space was actually landed on over the total number of rolls

taken by all players in all rounds. (The total number of rolls will be greater than the total number of tally marks, because the rolls that do not end exactly on the finish space are not recorded on the game board.)
4. To record the experimental probabilities, have students write the probability, in numeric form (12/293), in each space on the extra game board.
5. Once all groups have done this, spend some time comparing and contrasting the experimental data from group to group. Which spaces are very similar in the number of times they were landed on? Which spaces have very different results in each group?

Part Two
1. Distribute the student page for *Part Two* and one set of the special instructions cards to each group.
2. Go over the challenge and be sure that all groups understand the rules.
3. Allow time for students to come up with solutions for each of the challenges.
4. Share the solutions among the groups. If all groups solved any solution in the same way, try to find additional possible solutions. Each challenge can be solved in more than one way.

Part Three
1. Distribute the student page for *Part Three*.
2. Allow groups time to play the modified versions of their games and collect data.
3. Have a time of discussion and sharing where groups compare their results and identify whether or not they were successful in predicting the outcome of the modified game board.

Part Four
1. Distribute the student page for *Part Four*.
2. Allow groups time to determine their game board modifications and to collect data on the modified boards.
3. Once groups have had time to analyze their data, have them share how they modified the board and whether or not the modification produced the desired results.

Connecting Learning
Part One
1. What was the average number of rolls it took for you to complete the game? How did this compare to the group average?
2. How do your group's data compare to the data of the other groups?
3. What space on your game board was landed on most often? What was the experimental probability of landing there?

4. How did the experimental probabilities for specific spaces on the game board compare from group to group?
5. For which spaces was the experimental probability about the same from group to group? Why might this be?
6. For which spaces were there great differences in the experimental probability from group to group? Why might this be?

Part Two
1. How did you change the game so that it is possible to win it in three rolls? Was this the only way to make this change?
2. How did you increase the minimum number of rolls needed to complete the game? How else could you have accomplished this?
3. How did you increase the maximum number of rolls needed to complete the game? What else could you have done?
4. What did you do to decrease the maximum number of rolls to complete the game?
5. How did you change the minimum and maximum number of rolls possible? Did you choose to increase or decrease this value?
6. How do your solutions compare to those of other groups?

Part Three
1. What modification to the game board did you test?
2. How did you think this new game board would affect the average number of rolls needed to complete the game?
3. Do your data support your prediction? Why or why not?

Part Four
1. Describe the game board you used for this part of the activity.
2. Were you trying to make the average number of rolls go up or down?
3. Was your modified board successful? How do you know?
4. After playing the game, what changes would you make to your game board design? Why?

Extension
Have students design their own adaptation pieces to use on the game board and try them out.

Solutions
In order for any of the pieces to affect the minimum number of moves possible, they must be placed on spaces 6, 12, or 18. The reason for this is that in order to finish the game in the minimum number of moves, a six must be rolled every time. If a six is rolled every time, the only spaces that will be landed on are 6, 12, 18, and 24 (the finish). The adaptation pieces will all affect the maximum number of moves almost anywhere on the board. This is because the maximum number of moves is achieved by rolling a one every time, which means that every space is landed on.

The table that follows tells some of the ways that the maximum and minimum number of rolls can be affected by each of the adaptation pieces.

Piece	Result
Lose a turn (add one roll to your total)	**Minimum:** Increases to five if put on the sixth, 12th, or 18th space. **Maximum:** Increases to to 25 when put anywhere.
Double your next roll	**Minimum:** Decreases to three if put on the sixth, 12th, or 18th space. **Maximum:** Decreases to 23 when put anywhere but the 23rd space.
Move ahead six spaces	**Minimum:** Decreases to three if put on the sixth, 12th, or 18th space. **Maximum:** Decreases to 18 when put anywhere in the first 18 spaces.
Go back three spaces**	**Minimum:** Increases to five if put on the sixth, 12th, or 18th space. **Maximum:** Increases to 27 when put on any space.
Add four to your next roll	**Minimum:** Does not change. **Maximum:** Decreases to 20 when put in the first 20 spaces.

** This space can only penalize a player once per round. This prevents the theoretically possible loop that would occur if a one was rolled every time.

Probability on a Roll, Again

Key Questions

1. What is the experimental probability of landing on each space on the game board?
2. How can you use the adaptation pieces to change the average number of rolls it takes to complete the game?

Learning Goals

Students will:

- determine the experimental probability of landing on each space of the game board,
- use adaptation pieces to modify the game board in specific ways,
- determine the average number of rolls it takes to complete the modified game, and
- use two of the adaptation pieces to make a game board that takes more (or fewer) rolls on average to complete than the original board.

Probability on a Roll, Again

Challenge: Discover which spaces on the game board are landed on most often.

Rules:
- Play the game at least 10 times.
- Record the total number of rolls it takes each player in the table. Find the individual and group averages.
- Use the second game board to record tally marks. Each time someone lands on a space, make one mark.
- If your final roll puts you on the *Finish* space, make a tally mark in that space.
- If your final roll takes you past the *Finish* space, do not make a tally mark in that space.

PART 1

Number of Rolls

	Name:	Name:	Name:	Name:
Round 1				
Round 2				
Round 3				
Round 4				
Round 5				
Round 6				
Round 7				
Round 8				
Round 9				
Round 10				
Average				

Group Average

Probability Roll Again

on a Roll

Adaptation Pieces

Copy this page onto card stock and cut it into strips. Each group needs one strip.

Probability on a Roll, Again

Challenge: Use the adaptation pieces to change the game in specific ways.

Rules:
- You may only use one adaptation piece to solve each challenge.
- Each adaptation piece may only affect each player once each round. For example, if you landed on a space that read "Move back three spaces," did so, and then rolled a three on your next turn, you would not move back three spaces for a second time.

PART2

1. Change the game so that it is possible to finish in three rolls. Which adaptation piece do you need? Where does it need to go?

2. Increase the minimum number of rolls needed to complete the game. Which adaptation piece do you need? Where does it need to go?

3. Increase the maximum number of rolls needed to complete the game. Which adaptation piece do you need? Where does it need to go?

4. Decrease the maximum number of rolls needed to complete the game. Which adaptation piece do you need? Where does it need to go?

5. Change the minimum number of rolls possible (increase or decrease). Where does the adaptation piece need to go? Why?

6. Change the maximum number of rolls possible (increase or decrease). Where does the adaptation piece need to go? Why?

Probability on a Roll, Again

Challenge: Predict how a modified game board will affect the average number of rolls. Test your prediction to see if you are correct.

PART 3

1. Select one of the game boards that you developed in *Part Two* and describe it here.

2. How do you think the average number of rolls it will take to complete this game board will compare to the average for the original game board? Will it be higher, lower, or the same? Why?

3. In your group, play several times using the modified game board. Record the number of rolls it takes you to complete the game. Find the averages for each player and the combined group average.

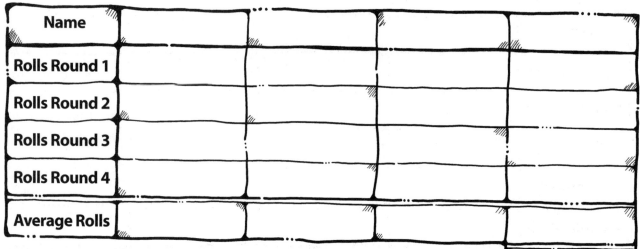

Name				
Rolls Round 1				
Rolls Round 2				
Rolls Round 3				
Rolls Round 4				
Average Rolls				

4. Make a bar graph showing each player's averages from this game and their averages from *Part One*. Was your prediction correct? Why or why not?

Group Average

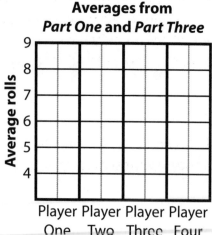

Averages from
Part One* and *Part Three

Average rolls

9
8
7
6
5
4

Player   Player   Player   Player
One   Two   Three   Four

Probability on a Roll, Again

Challenge: Change the game so that the average number of rolls it takes to finish goes up or down by at least one when compared to *Part One* data.

Rules: • Use only two of the adaption pieces.
• Each special space can only affect each player once each round. For example, if you land on the "Go back three spaces" piece for a second time, you do not move back again.

1. Which two pieces will you use?

2. Where will you put the pieces? Be specific.

3. Is your goal to increase or decrease the average number of rolls?

4. Play the game with your group several times. Keep track of the number of rolls it takes you to finish each round. Find the individual and group averages.

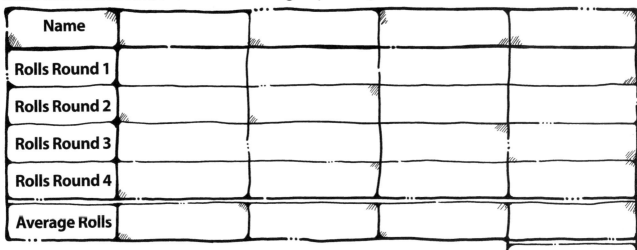

Name				
Rolls Round 1				
Rolls Round 2				
Rolls Round 3				
Rolls Round 4				
Average Rolls				

Group Average

5. Make a bar graph comparing each player's averages from this game to the averages they had in *Part One* of this activity. Were you successful in creating a game that had either higher or lower averages? Why or why not?

6. If you could make any changes to your game board design, what would they be?

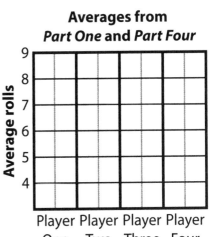

Averages from Part One and Part Four

Average rolls
9
8
7
6
5
4

Player One   Player Two   Player Three   Player Four

Connecting Learning

Part One

1. What was the average number of rolls it took for you to complete the game? How did this compare to the group average?

2. How do your group's data compare to the data of the other groups?

3. What space on your game board was landed on most often? What was the experimental probability of landing there?

4. How did the experimental probabilities for specific spaces on the game board compare from group to group?

5. For which spaces was the experimental probability about the same from group to group? Why might this be?

6. For which spaces were there great differences in the experimental probability from group to group? Why might this be?

Connecting Learning

Part Two

1. How did you change the game so that it is possible to win it in three rolls? Was this the only way to make this change?

2. How did you increase the minimum number of rolls needed to complete the game? How else could you have accomplished this?

3. How did you increase the maximum number of rolls needed to complete the game? What else could you have done?

4. What did you do to decrease the maximum number of rolls to complete the game?

5. How did you change the minimum and maximum number of rolls possible? Did you choose to increase or decrease this value?

6. How do your solutions compare to those of other groups?

Connecting Learning

Part Three

1. What modification to the game board did you test?

2. How did you think this new game board would affect the average number of rolls needed to complete the game?

3. Do your data support your prediction? Why or why not?

Part Four

1. Describe the game board you used for this part of the activity.

2. Were you trying to make the average number of rolls go up or down?

3. Was your modified board successful? How do you know?

4. After playing the game, what changes would you make to your game board design? Why?

CATCH AND RELEASE

Topic
Sampling

Key Question
How can scientists estimate the populations of wild animals if the animals are never all together to count?

Learning Goals
Students will:
- use random sampling to get a ratio of marked counters to all counters in a bag,
- use proportional reasoning to determine the total number of counters in a bag given how many in the sample were marked, and
- recognize applications for probability.

Guiding Documents
Project 2061 Benchmarks
- *The larger a well-chosen sample is, the more accurately it is likely to represent the whole, but there are many ways of choosing a sample that can make it unrepresentative of the whole.*
- *Use calculators to compare amounts proportionally.*

Common Core State Standards for Mathematics*
- *Make sense of problems and persevere in solving them. (MP.1)*
- *Reason abstractly and quantitatively. (MP.2)*
- *Construct viable arguments and critique the reasoning of others. (MP.3)*
- *Model with mathematics. (MP.4)*
- *Use appropriate tools strategically. (MP.5)*
- *Use random sampling to draw inferences about a population. (7.SP.A)*

Math
Estimation
Data analysis
    sampling
    probability
Proportional reasoning
    ratios
Problem solving

Integrated Processes
Observing
Predicting
Collecting and recording data
Interpreting data
Analyzing
Applying

Materials
For each group:
    one paper lunch bag
    counters in two colors (see *Management 1*)

For each student:
    student page

Background Information
There are two principal ways of gathering quantitative data—by census or by sample. In a census, every organism, object, event, etc., is counted. Since it is usually impractical or impossible to count every element, the preferred technique is sampling. One of the most frequently used methods of sampling is random sampling. In random sampling, each element has an equal chance of appearing in the sample.

To count the population of a species of animal in the wild, naturalists use a method of sampling where animals are caught, tagged, and released. A certain number of animals are caught and tagged to identify them as having been caught. The tagged animals are then released. As animals are caught subsequently, the ratio of tagged animals caught to all animals caught provides a way to predict the population of the animals in the wild. If 10% of the animals caught are tagged, it is probable that tagged animals are 10% of the whole population.

To simulate a catch and release survey, place from 25 to 50 counters of the same color into the bag to represent the total population. Tagging is simulated by grabbing 10 of the counters and replacing them with counters of a different color. The different color counters represent tagged animals. As samples are taken out of the bag, the ratio of different colored counters to total counters is proportional to the actual number of different colored counters to total counters in the bag. The more samples taken, the more accurately the average ratio will match the contents of the bag.

Management

1. Any uniform manipulatives that come in multiple colors can be used as counters. Teddy Bear Counters, Astronaut Counters, and Area Tiles are a few examples of items that work well. All of these manipulatives can be ordered from AIMS.

2. Prepare a bag of counters for each group that contains 25-50 counters of one color. Each group needs an additional 10 counters in another color that are outside the bag.

Procedure

1. Ask the *Key Question* and discuss students' ideas and suggestions.

2. Have students get into groups and distribute the materials.

3. Instruct the students to follow the directions on the student page to simulate the catch, tag, and release.

4. Have the students complete the instructions by taking five samples from the bag and recording the results. Have the students return the counters to the bag and shake it to randomize the counters between draws.

5. Using their data, have students make a proportion and predict the total number of counters in the bag.

6. Have the students count the counters in the bag and see how closely their predictions came to the actual contents.

7. Discuss the results of the investigation.

Connecting Learning

1. Which ratio is closer to what was really in the bag—the total ratio or a sample ratio? [The larger the sample, the more accurate it tends to be, so the total ratio tends to be closer to what is in the bag.]

2. Why did you shake the bag between samples? [to mix them up, random]

3. How well did your proportion predict the number of counters in the bag?

4. Was using this random sample method a good way to get an accurate estimate of what is in the bag? Explain.

5. How could you apply this random sampling method to count a large mobile population of animals?

6. What other ways can you think of using a sampling method?

CATCH AND RELEASE

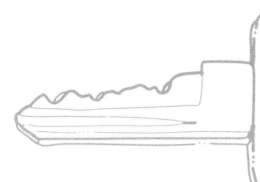

Key Question

How can scientists estimate the populations of wild animals if the animals are never all together to count?

Learning Goals

Students will:

- use random sampling to get a ratio of marked counters to all counters in a bag,
- use proportional reasoning to determine the total number of counters in a bag given how many in the sample were marked, and
- recognize applications for probability.

CATCH AND RELEASE

To count the population of a species of animal in the wild, naturalists use a method where a number of animals are caught and tagged. The tagged animals are released back into the wild. As animals are caught later, the ratio of the tagged animals to total animals caught provides a way to predict the population of the animals in the wild.

Simulate this method of sampling to determine the number of counters in your bag.
1. Without looking, reach into your bag and catch 10 counters.
2. Exchange these counters for counters of another color. The new color counters represent animals that have been caught and tagged.
3. Put the tagged counters into the bag and shake the bag.
4. Without looking into the bag, randomly draw five to 10 counters.
5. Record the number of "tagged" counters and the total number of counters grabbed, along with the ratios.
6. Replace the counters in the bag and shake the bag between samples. Take and record four more samples to complete the chart.

Sample	"Tagged" Counters	Total Counters	Ratio $\frac{\text{"Tagged"}}{\text{Total}}$	Decimal Equivalent	Percent "Tagged"
One					
Two					
Three					
Four					
Five					
Total					

Make a proportion using the totals and predict the number of counters in the bag.

Sample Totals

"Tagged" Counters

Total Counters

=

Bag Totals

10
"Tagged" Counters

Total Counters

How well did your prediction match the actual number of counters in the bag?

CATCH AND RELEASE

Connecting Learning

1. Which ratio is closer to what was really in the bag—the total ratio or a sample ratio?

2. Why did you shake the bag between samples?

3. How well did your proportion predict the number of counters in the bag?

4. Was using this random sample method a good way to get an accurate estimate of what is in the bag? Explain.

5. How could you apply this random sampling method to count a large mobile population of animals?

6. What other ways can you think of using a sampling method?

Topic
Probability

Key Question
What is the probability that a basketball player will make a free throw?

Learning Goals
Students will:
- find the fraction of free throws made by each of the five starters for a basketball team,
- use the fraction to predict how each player will perform at the free throw line,
- record probabilities as decimals and percents, and
- observe that the sum of the probabilities of a player hitting a free throw and missing a free throw is 1.

Guiding Documents
Project 2061 Benchmarks
- *How probability is estimated depends on what is known about the situation. Estimates can be based on data from similar conditions in the past or on the assumption that all the possibilities are known.*
- *Probabilities are ratios that can be expressed as fractions, percentages, or odds.*

Common Core State Standards for Mathematics*
- *Make sense of problems and persevere in solving them. (MP.1)*
- *Reason abstractly and quantitatively. (MP.2)*
- *Construct viable arguments and critique the reasoning of others. (MP.3)*
- *Model with mathematics. (MP.4)*
- *Use appropriate tools strategically. (MP.5)*
- *Investigate chance processes and develop, use, and evaluate probability models. (7.SP.C)*

Math
Data analysis
    probability
    mutually exclusive events
Proportional reasoning

Integrated Processes
Observing
Comparing and contrasting
Collecting and recording data
Analyzing
Applying

Materials
Student pages

Background Information
Probability is a measure of uncertainty; it is a measure of likelihood of an uncertain event; it is a measure with values between 0 and 1. These are key ideas that we want students to come to understand.

There are at least three strands of earlier learning with which students should make connections as they begin to develop an understanding of probability. These are:
- collecting and organizing data and making inferences from data,
- finding ratios and using them in proportional reasoning situations, and
- understanding measurement as the assigning of a number to an attribute of an object or event.

The probability of an outcome of an action can sometimes be determined theoretically. This is the case in rolling a die or flipping a coin. More often in real-world situations there is no way to determine the probability of an outcome except experimentally, that is, by repeatedly performing an action and keeping a record of the outcomes. When we do that, what we are finding is an estimate of the probability of that outcome. The more times we perform the action, the better will be the estimate. This is the case with the situation in this activity. The only way to find the probability that a player will make a free throw is by looking at her performance over the first 14 games. What we are finding is an estimate of the probability that a player will make a free throw.

Management
1. The activity is best done in small groups, but with whole group interaction as well. While students should complete the charts themselves, the questions are best discussed first in small groups and then posed to the entire class.
2. Each student needs his or her own copy of the student pages.

Procedure
1. Ask students to read the opening paragraph and look at the first chart. Talk with students about the different statistics that are kept on each player in a basketball game.

2. Invite students to discuss in groups or with a partner the two questions about which players on the team were the best or worst free throw shooters. Bring the discussion back to the entire group. Students may simply be looking to see which players attempted the greatest or the least number of shots.

3. Direct students to look at the chart and order the players from most likely to least likely to make a free throw. Ask them to compare and give some reasons for their ordering.

4. Have students read the paragraph at the top of the second student page. The point of this paragraph is simply to help students see that probability is a measure of something. We want them to understand that probability is the measure of uncertainty. Shooting a free throw is an uncertain event. The player may make it or miss it. Probability gives us a way to assign a number to such an uncertain event, just like the bathroom scale gives us a way to assign a number to our weight.

5. Ask students to go back to the chart of free throws and have them use it to complete the next table, where they find the ratio of free throws made to free throws attempted and express it as a fraction, a decimal, and a percent. Tell them that these are the numbers that tell us the probability of making a free throw for the members of the team.

6. Using the chart showing the fractions of free throws made by each player, have students again order the players from most likely to least likely to make a free throw. Ask them to compare their ordering this time with the previous one.

7. The first paragraph on the third student page is intended to reinforce the idea that probability is a measure of uncertainty and to remind students about how these probability measures were found and what they mean. It should be noted that saying that the probability of a player hitting a free throw is .67, is the same thing as saying that she is a 67% free throw shooter.

8. Have students read the second paragraph on the third student sheet. Ask them to discuss in groups who they think the coach would choose and why. Again, bring the discussion back to the entire group.

9. Direct students to complete the final chart that has some duplicate information, but introduces some new columns showing the number of free throws missed.

10. Explore with students the relationship between hits and misses. They should notice that the sum of hits and misses is equal to the total number of attempts. More importantly, they should notice that $P(\text{Hit}) + P(\text{Miss}) = 1$.

Connecting Learning
1. How does knowing the probabilities of the players making free throws help in ordering them according to their abilities?
2. Is the player who attempts the most free throws necessarily the best free throw shooter? Why or why not?
3. Is the player who attempts the fewest free throws in a game necessarily the worst free throw shooter? Why or why not?
4. If you had the chart for the team at the end of the season, after they have played six more games, do you suppose the probabilities would be exactly the same? Why or why not?
5. What are the two possible outcomes when a player shoots a free throw? Does it make sense then that $P(\text{Hit}) + P(\text{Miss}) = 1$?

Solutions

Player	FTA	FT	$\frac{FT}{FTA}$	Decimal	Percent
Taylor	78	51	51/78	0.65	65%
Jasmine	62	36	36/62	0.58	58%
Dominique	41	31	31/41	0.76	76%
Natalie	25	17	17/25	0.68	68%
Olivia	19	7	7/19	0.37	37%

The players in order from most likely to make a free throw to least likely are as follows:
Dominique, Natalie, Taylor, Jasmine, Olivia

Player	FTA	FT	FTM	$\frac{FT}{FTA}$	$\frac{FTM}{FTA}$
Taylor	78	51	27	51/78	27/78
Jasmine	62	36	26	36/62	26/62
Dominique	41	31	10	31/41	10/41
Natalie	25	17	8	17/25	8/25
Olivia	19	7	12	7/19	12/19

Key Question

What is the probability that a basketball player will make a free throw?

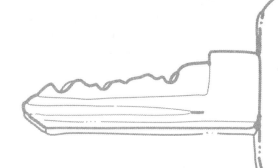

Learning Goals

Students will:

- find the fraction of free throws made by each of the five starters for a basketball team,

- use the fraction to predict how each player will perform at the free throw line,

- record probabilities as decimals and percents, and

- observe that the sum of the probabilities of a player hitting a free throw and missing a free throw is 1.

Player Stat Sheet

Coach Brown is concerned that the players on her basketball team at North Side High School don't fully understand how important good free throw shooting is to winning ball games. To help them focus on how they are doing, she has posted free throw shooting statistics for her starting five players on the wall in the gym. The following chart shows the statistics for these players for the first 14 games of the year.

Taylor FT-FTA	Jasmine FT-FTA	Dominique FT-FTA	Natalie FT-FTA	Olivia FT-FTA	Opponent
1 – 4	1 – 2	2 – 2	0 – 0	2 – 2	Elwood
5 – 7	5 – 11	4 – 5	0 – 0	0 – 0	Bosse
1 – 2	6 – 8	0 – 0	1 – 2	3 – 3	Fairfield
5 – 5	3 – 4	5 – 10	2 – 2	0 – 4	Forest Park
6 – 8	0 – 2	2 – 2	2 – 2	0 – 0	West Central
2 – 2	0 – 0	2 – 2	0 – 0	0 – 0	Warren
7 – 10	3 – 4	0 – 0	3 – 5	0 – 0	Taylor
2 – 2	5 – 7	4 – 5	4 – 5	0 – 0	Tell City
3 – 6	3 – 4	2 – 2	2 – 2	0 – 0	Bend
5 – 9	2 – 3	6 – 7	0 – 0	1 – 2	Ross
3 – 6	2 – 5	2 – 2	0 – 1	1 – 4	Newton
4 – 6	3 – 7	0 – 2	1 – 2	0 – 4	Spencer
3 – 5	2 – 2	0 – 0	2 – 4	0 – 0	Valley Oaks
4 – 6	1 – 3	2 – 2	0 – 0	0 – 0	Knox Spring

FT = successful free throws, FTA = total free throws attempted

Through the first 14 games, who made the most free throws? Was she the best free throw shooter on the team? Why or why not?

Through the first 14 games, who made the fewest free throws? Was she the worst free throw shooter on the team? Why or why not?

In the blanks, write the names of the five players in order from most likely to least likely to make a free throw. How did you decide the ordering?

_____ _____ _____ _____ _____

Most Likely   Least Likely

From Data to Probability

Writing the players names in order from most likely to least likely to hit a free throw is similar to putting them in order from heaviest to lightest or tallest to shortest without weighing them or measuring their heights. It would be easier and more accurate if we knew the measurements for each player. Just as it is possible to measure a player's height, there is a way to measure the chance or likelihood that a player will make a free throw. That measurement is called *probability*. The probability that a player will make a free throw is a number that gives us a measurement of a player's chance of making a free throw. The probability can be found by dividing the number of free throws made by the number of free throws attempted.

Go back to the free throw chart and find the total number of free throws each player attempted and the number that each player made. Put these numbers in the following table. Then complete the table by writing the fraction of *free throws made* over *free throws attempted* (FT/FTA), and write the fraction as a decimal and as a percent. These numbers are the probability for each player of making a free throw.

Player's Name	Free Throws Attempted (FTA)	Successful Free Throws (FT)	$\dfrac{FT}{FTA}$	Decimal	Percent
Taylor					
Jasmine					
Dominique					
Natalie					
Olivia					

Order the players again from most likely to make a free throw to least likely. This time, use their probability of making a free throw as the basis for the ordering.

_____ _____ _____ _____ _____

Most Likely   Least Likely

Was this ordering different from the first one? Why or why not?

From Probability to Expectation

The player most likely to make a free throw is the one with the highest probability. Just as we know that a player who is 6'5" is taller than one who is 5'11", we know that a player with a probability of .67 of hitting a free throw is more likely to hit a shot than the player with a probability of .45. Sometimes the probability of hitting free throws is expressed as a percent. Rather than saying that the probability is .67 that Taylor will hit a free throw, we could say that she is a 67% free throw shooter.

Suppose that at the end of the next game, with two seconds left on the clock, North Side is behind by one point. The referee's whistle blows and he calls a foul on the opposing team. It takes the referee a couple of seconds to point to the North Side player who was fouled. Who do suppose coach Brown would most like to be shooting free throws in this situation? Why?

If you know the number of free throws attempted by a player and you know the number of free throws that she made, how can you find the number that she missed? If you know the probability that a player will hit a free throw, how can you find the probability that she will miss?

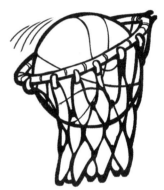

Complete the following chart using the data from the first two pages.

Player's Name	Free Throws Attempted (FTA)	Successful Free Throws (FT)	Free Throws Missed (FTM)	$\dfrac{FT}{FTA}$	$\dfrac{FTM}{FTA}$
Taylor					
Jasmine					
Dominique					
Natalie					
Olivia					

How does the probability of a player missing a free throw compare to the probability of hitting a free throw?

Connecting Learning

1. How does knowing the probabilities of the players making free throws help in ordering them according to their abilities?

2. Is the player who attempts the most free throws necessarily the best free throw shooter? Why or why not?

3. Is the player who attempts the fewest free throws in a game necessarily the worst free throw shooter? Why or why not?

4. If you had the chart for the team at the end of the season, after they have played six more games, do you suppose the probabilities would be exactly the same? Why or why not?

5. What are the two possible outcomes when a player shoots a free throw? Does it make sense then that P(Hit) + P(Miss) = 1?

Metric Measuring Tapes

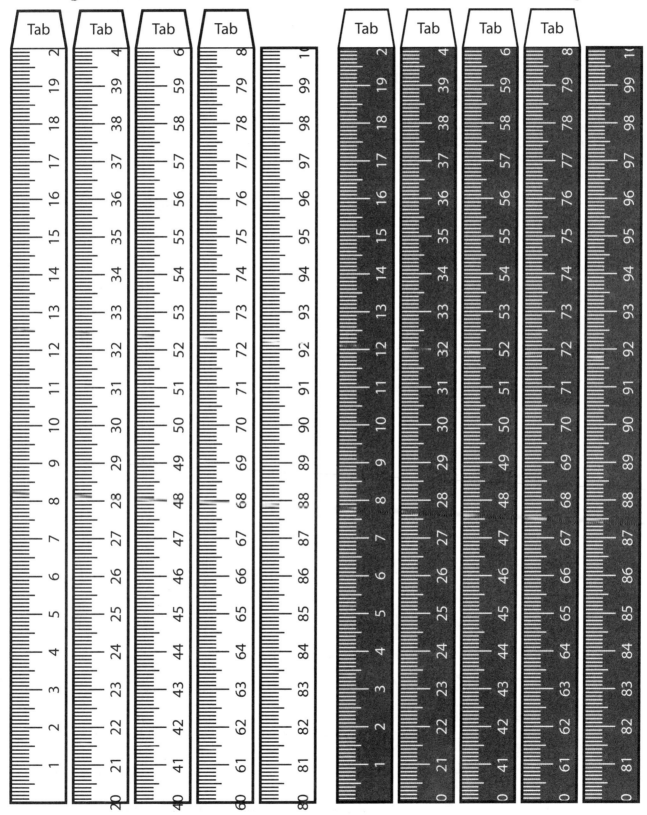

The AIMS Program

AIMS is the acronym for "Activities Integrating Mathematics and Science." Such integration enriches learning and makes it meaningful and holistic. AIMS began as a project of Fresno Pacific University to integrate the study of mathematics and science in grades K-9, but has since expanded to include language arts, social studies, and other disciplines.

AIMS is a continuing program of the non-profit AIMS Education Foundation. It had its inception in a National Science Foundation funded program whose purpose was to explore the effectiveness of integrating mathematics and science. The project directors, in cooperation with 80 elementary classroom teachers, devoted two years to a thorough field-testing of the results and implications of integration.

The approach met with such positive results that the decision was made to launch a program to create instructional materials incorporating this concept. Despite the fact that thoughtful educators have long recommended an integrative approach, very little appropriate material was available in 1981 when the project began. A series of writing projects ensued, and today the AIMS Education Foundation is committed to continuing the creation of new integrated activities on a permanent basis.

The AIMS program is funded through the sale of books, products, and professional-development workshops, and through proceeds from the Foundation's endowment. All net income from programs and products flows into a trust fund administered by the AIMS Education Foundation. Use of these funds is restricted to support of research, development, and publication of new materials. Writers donate all their rights to the Foundation to support its ongoing program. No royalties are paid to the writers.

The rationale for integration lies in the fact that science, mathematics, language arts, social studies, etc., are integrally interwoven in the real world, from which it follows that they should be similarly treated in the classroom where students are being prepared to live in that world. Teachers who use the AIMS program give enthusiastic endorsement to the effectiveness of this approach.

Science encompasses the art of questioning, investigating, hypothesizing, discovering, and communicating. Mathematics is a language that provides clarity, objectivity, and understanding. The language arts provide us with powerful tools of communication. Many of the major contemporary societal issues stem from advancements in science and must be studied in the context of the social sciences. Therefore, it is timely that all of us take seriously a more holistic method of educating our students. This goal motivates all who are associated with the AIMS Program. We invite you to join us in this effort.

Meaningful integration of knowledge is a major recommendation coming from the nation's professional science and mathematics associations. The American Association for the Advancement of Science in *Science for All Americans* strongly recommends the integration of mathematics, science, and technology. The National Council of Teachers of Mathematics places strong emphasis on applications of mathematics found in science investigations. AIMS is fully aligned with these recommendations.

Extensive field testing of AIMS investigations confirms these beneficial results:

1. Mathematics becomes more meaningful, hence more useful, when it is applied to situations that interest students.
2. The extent to which science is studied and understood is increased when mathematics and science are integrated.
3. There is improved quality of learning and retention, supporting the thesis that learning which is meaningful and relevant is more effective.
4. Motivation and involvement are increased dramatically as students investigate real-world situations and participate actively in the process.

We invite you to become part of this classroom teacher movement by using an integrated approach to learning and sharing any suggestions you may have. The AIMS Program welcomes you!

Get the Most From Your Hands-on Teaching

When you host an AIMS workshop for elementary and middle school educators, you will know your teachers are receiving effective, usable training they can apply in their classrooms immediately.

AIMS Workshops are Designed for Teachers
- Hands-on activities
- Correlated to your state standards
- Address key topic areas, including math content, science content, and process skills
- Provide practice of activity-based teaching
- Address classroom management issues and higher-order thinking skills
- Include $50 of materials for each participant
- Offer optional college (graduate-level) credits

AIMS Workshops Fit District/Administrative Needs
- Flexible scheduling and grade-span options
- Customized workshops meet specific schedule, topic, state standards, and grade-span needs
- Sustained staff development can be scheduled throughout the school year
- Eligible for funding under the Title I and Title II sections of No Child Left Behind
- Affordable professional development—consecutive-day workshops offer considerable savings

Call us to explore an AIMS workshop
1.888.733.2467

Online and Correspondence Courses
AIMS offers online and correspondence courses on many of our books through a partnership with Fresno Pacific University.
- Study at your own pace and schedule
- Earn graduate-level college credits

© 2012 AIMS Education Foundation

AIMS Program Publications

Actions With Fractions, 4-9
The Amazing Circle, 4-9
Awesome Addition and Super Subtraction, 2-3
Bats Incredible! 2-4
Brick Layers II, 4-9
The Budding Botanist, 3-6
Chemistry Matters, 5-7
Concerning Critters: Adaptations &
    Interdependence, 3-5
Counting on Coins, K-2
Cycles of Knowing and Growing, 1-3
Crazy About Cotton, 3-7
Critters, 2-5
Earth Book, 6-9
Earth Explorations, 2-3
Earth, Moon, and Sun, 3-5
Earth Rocks! 4-5
Electrical Connections, 4-9
Energy Explorations: Sound, Light, and Heat, 3-5
Exploring Environments, K-6
Fabulous Fractions, 3-6
Fall Into Math and Science*, K-1
Field Detectives, 3-6
Floaters and Sinkers, 5-9
From Head to Toe, 5-9
Getting Into Geometry, K-1
Glide Into Winter With Math and Science*, K-1
Gravity Rules! 5-12
Hardhatting in a Geo-World, 3-5
Historical Connections in Mathematics, Vol. I, 5-9
Historical Connections in Mathematics, Vol. II, 5-9
Historical Connections in Mathematics, Vol. III, 5-9
It's About Time, K-2
It Must Be A Bird, Pre-K-2
Jaw Breakers and Heart Thumpers, 3-5
Looking at Geometry, 6-9
Looking at Lines, 6-9
Machine Shop, 5-9
Magnificent Microworld Adventures, 6-9
Marvelous Multiplication and Dazzling Division, 4-5
Math + Science, A Solution, 5-9
Mathematicians are People, Too
Mathematicians are People, Too, Vol. II
Mostly Magnets, 3-6
Movie Math Mania, 6-9
Multiplication the Algebra Way, 6-8
Out of This World, 4-8
Paper Square Geometry:
    The Mathematics of Origami, 5-12
Popping With Power, 3-5
Positive vs. Negative, 6-9
Primarily Bears*, K-6
Primarily Critters, K-2
Primarily Magnets, K-2

Primarily Physics: Investigations in Sound, Light,
    and Heat Energy, K-2
Primarily Plants, K-3
Primarily Weather, K-3
Probing Space, 3-5
Problem Solving: Just for the Fun of It! 4-9
Problem Solving: Just for the Fun of It! Book Two, 4-9
Proportional Reasoning, 6-9
Puzzle Play, 4-8
Ray's Reflections, 4-8
Sensational Springtime, K-2
Sense-able Science, K-1
Shapes, Solids, and More: Concepts in Geometry, 2-3
Simply Machines, 3-5
The Sky's the Limit, 5-9
Soap Films and Bubbles, 4-9
Solve It! K-1: Problem-Solving Strategies, K-1
Solve It! 2nd: Problem-Solving Strategies, 2
Solve It! 3rd: Problem-Solving Strategies, 3
Solve It! 4th: Problem-Solving Strategies, 4
Solve It! 5th: Problem-Solving Strategies, 5
Solving Equations: A Conceptual Approach, 6-9
Spatial Visualization, 4-9
Spills and Ripples, 5-12
Spring Into Math and Science*, K-1
Statistics and Probability, 6-9
Through the Eyes of the Explorers, 5-9
Under Construction, K-2
Water, Precious Water, 4-6
Weather Sense: Temperature, Air Pressure, and
    Wind, 4-5
Weather Sense: Moisture, 4-5
What on Earth? K-1
What's Next, Volume 1, 4-12
What's Next, Volume 2, 4-12
What's Next, Volume 3, 4-12
Winter Wonders, K-2

Essential Math
Area Formulas for Parallelograms, Triangles, and
    Trapezoids, 6-8
Circumference and Area of Circles, 5-7
Effects of Changing Lengths, 6-8
Measurement of Prisms, Pyramids, Cylinders, and
    Cones, 6-8
Measurement of Rectangular Solids, 5-7
Perimeter and Area of Rectangles, 4-6
The Pythagorean Relationship, 6-8
Solving Equations by Working Backwards, 7

*   Spanish supplements are available for these books. They
    are only available as downloads from the AIMS website.
    The supplements contain only the student pages in
    Spanish; you will need the English version of the book for
    the teacher's text.

For further information, contact:
AIMS Education Foundation • 1595 S. Chestnut Ave. • Fresno, California 93702
www.aimsedu.org • 559.255.6396 (fax) • 888.733.2467 (toll free)

Duplication Rights

No part of any AIMS publication—digital or otherwise—may be reproduced or transmitted in any form or by any means—except as noted below.